The OLD HOUSE BOOK of
BARN PLANS

The OLD HOUSE BOOK of
BARN PLANS

Richard Rawson

A Sterling/Main Street Book
Sterling Publishing Co., Inc.   New York

Library of Congress Cataloging-in-Publication Data

Rawson, Richard, 1950–
    [Old barn plans]
    The Old house book of barn plans / Richard Rawson.
        p.   cm.
    Previous published under title: Old barn plans.
    "A Sterling/Main Street book.
    1. Barns—United States—Designs and plans.   I. Title.
NA8230.R38   1990   90-9943
728′.922′0223—dc20   CIP

10   9   8   7   6   5   4   3   2   1

A Sterling / Main Street Book

Published 1990 by Sterling Publishing Company, Inc.
387 Park Avenue South, New York, N.Y. 10016
Originally published by the Mayflower Press as
Old Barn Plans © 1979 by The Main Street Press
Distributed in Canada by Sterling Publishing
% Canadian Manda Group, P.O. Box 920, Station U
Toronto, Ontario, Canada M8Z 5P9
Distributed in Great Britain and Europe by Cassell PLC
Villiers House, 41/47 Strand, London WC2N 5JE, England
Distributed in Australia by Capricorn Ltd.
P.O. Box 665, Lane Cove, NSW 2066

Sterling ISBN 0-8069-7417-6

contents

preface

Since most of us are now citified or at least sub-urban, we've probably never set foot in a barn. Maybe the closest we've come to one is while driving on a superhighway that slices through farmland. We might have been suitably impressed by its scale and solitude on the horizon and perhaps wondered why half its siding had fallen off while what remained appeared to be reasonably sound. In all likelihood the siding didn't just weather away, but rather was removed by a scavenger eager to peddle rusticity on the open market. We may even have some of it nailed to our own living room ceiling or walls.

It's not hard to imagine how old barns come apart, but we probably do not know much about how they were put together, or when, or even why. *Old Barn Plans* is about barns of the past and barns that are still standing. It is filled with floor plans and perspectives of barns built in the United States and Canada during the three centuries between the first coastal settlements and the dawn of the 20th century. It depicts the evolution of an ancient form from its primitive state in the North American wilderness to a veritable living laboratory of plant and animal husbandry. *Old Barn Plans* provides indispensable information to anyone rebuilding an old barn, to anyone wishing to build a new barn along traditional lines, or to just the curious.

I wish to thank Alicia Stamm of the Historic American Buildings Survey, Mary Ison at the Library of Congress, and Vicky Williams at the Glenbow-Alberta Institute in Calgary, Alberta, for helping me to track down many of the illustrations in this book. Of equal importance is the scholarship of those on whose work the text of this book is based: C. A. Weslager, John Fitchen, Henry Glassie, Amos Long, Jr., Dolores Hayden, and Eric Arthur.

The illustrations on pages 16, 18-26, 32, 34-43, 50-53, 56-60, 78-95, and 100-104 are courtesy, Historic American Buildings Survey, Washington, D.C.; pages 149-154 courtesy, The Glenbow-Alberta Institute, Calgary, Alberta. Those on pages 136, 138-148 first appeared in *Louden Barn Plans*, Louden Machinery Company, 1917; and on pages 155-157, in (William) *Radford's Practical Barn Plans*, Radford Architectural Company, 1909. All other material is borrowed from the pages of *The American Agriculturist*, 1859-1890.

HARVEST

introduction

For the earliest European settlers in North America, wood was the most plentiful and exploitable of resources from which to build shelters. Nearly three centuries later this was still true for the pioneers moving into the last western frontiers of the United States and Canada. Virgin forests yielded to the firstcomers the raw material from which simple thatched or sod-covered hovels and lean-tos could be built. Often, a stockade-like shelter was constructed from a series of vertically-arranged logs. A proper cabin or house might not be built for several seasons or even years while the family struggled to insure its survival. Later arrivals, if they were lucky, were taken into the homes of neighbors until a cabin could be built.

What constituted proper pioneer housing in the 17th and 18th centuries depended on an immigrant's national origin. For example, the English adventurers and speculators who met with uneven but ultimate success at Jamestown in 1607, and the Pilgrims who sailed into Massachusetts Bay in the 1620s attempted to recreate the thatch- and clapboard-covered frame dwellings and barns they had left behind. This involved time-consuming and laborious efforts. Their architectural and cultural sophistication left them sorely unprepared for surviving in the wilderness. Had their tradition included the simple horizontal piling of logs one upon the other, perhaps fewer poor souls would have frozen to death for lack of shelter.

Clearing the land required a sharp axe, a strong back, and a will of steel. Often, before a barn could be built, a lean-to was set up to protect the livestock.

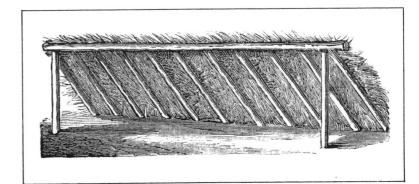

Native Americans saved some lives by demonstrating the use of tepees. Only when a toehold was established and basic farming began was there time to erect barns for crops and sheds for livestock.

In contrast to the English, the Swedish-Finnish settlers were initially more successful. Because of their later migration (the Swedes and Finns first arrived in the Delaware Valley in 1638, and the Germans began to settle in the 1680s), they benefitted from the firsthand accounts of returning Europeans. In addition, the Central European and Scandinavian traditions of log construction held them in good stead. Their log barns housed both livestock and feed and usually were larger and more fit for habitation than their humble dwellings. The methods and materials were so sensible and correct for the North American environment that the German homes and barns of southeastern Pennsylvania became the primary models for the farms that were rapidly spreading southward along Virginia's Great Valley, across the Blue Ridge, and northward into New York and Ontario.

Log houses and log barns were still being built in the western United States and in western Canada well into the 19th century.

New York's Hudson and Mohawk Valleys were bastions of Dutch influence by the end of the 17th century. Like their Scandinavian and German contemporaries to the south, the Dutch were quick to erect barns and slow to build houses. Their barns were easily the most sophisticated yet seen in the colonies. Using framing techniques common in Holland, Dutch farmers created a unique hybrid designed to accommodate North American crops, climate, geography, and building materials. The Dutch barn was built entirely of wood and contained a voluminous hayloft over a central threshing floor flanked by two aisles for livestock. The design flourished with little modification from New York across the St. Lawrence into Ontario, and in parts of New Jersey, especially in Bergen County and along the Millstone River near Princeton.

Increased demand for farm goods in the late 18th century fueled a boom in frame barn construction in the mid-Atlantic region. This was most evident among the Pennsylvania Germans. What appeared with increasing frequency in the early 1800s was a large two- and sometimes three-level structure built into a slope or bank that incorporated and advanced many of the developments found in the log barns, some of which had by now become quite large. Like its predecessor, the new Pennsylvania barn was a model of appropriateness in time and circumstance; it met the current need for a general purpose building to accommodate more stock and larger volumes of hay on increasingly productive farms. Migration to the west and

Fire has swept away thousands of barns. Here, farmers do what they can to save the hay and horses, but the Dutch barn is doomed.

south from the Pennsylvania region insured the rapid adoption of this and similar designs by farmers from the southern Appalachians to western Ontario in the mid-19th century, the popular magazine *The American Agriculturist* championed the Pennsylvania barn's utility as an indispensable labor-saving tool and offered its own floor plans to inspire subscribers.

The magazine was not so effusive about the English, or Yankee barn, "an institution known from one end of the country to the other—certainly throughout the older Northern States"—which was "unsightly, inconvenient, and poorly adapted to any use but that of storing hay and straw," functions which sufficed in England but ignored the need for housing livestock in a more hostile climate. By 1866, when this comment appeared, grain production in the Northeast was waning; dairy farming and other specialties were becoming more profitable pursuits. As a result, the magazine offered numerous remedies to farmers unfortunate enough to be strapped with an English barn, most of which involved transforming it into something approximating a Pennsylvania barn.

The pressures of the marketplace were not the only forces working to change the rural landscape in the 19th century. The intoxicating promise of democracy and the seemingly limitless availability of cheap land in both the United States and Canada helped generate a spirit of experimentation in architecture and agriculture. Social ferment also began to simmer as activists raised their

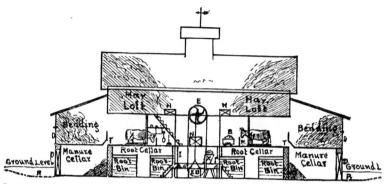

Inside a 19th-century barn.

voices for abolition, women's rights, and educational and penal reforms. Waves of irrepressible revivalism carried the shrill jeremiads of impending doom to slumbering sinners. The specter of moral decay first heralded in dismal English industrial towns reared its head in America.

Among the many utopian groups which professed soul-saving alternatives were the Shakers, a growing, self-contained society of perfectionists who, before their decline later in the century, would make a lasting contribution to architecture, furniture, and the decorative arts in North America. In response to the spiritual and physical pollution of an industrializing nation, adherents created a unique communal arrangement dedicated to God's work and worship; they were part of what they called a "living building" of earth, wood, body, and spirit.

Shaker settlements were not designed by architects, but by master builders who carried out construction projects at least as sophisticated as those

Until the late 19th century, a particularly sophisticated horse power device, *top,* was used to drive threshers and cutters and to lift heavy loads in the barn. *Bottom:* Steam power, in use early in the century, did not fully replace horses until the 1900s, by which time tractors were being run by internal combustion engines.

of their professional contemporaries. Shaker barns, like Shaker houses, were exceptionally spacious without appearing monumental. Even the great North Family barn in New Lebanon, New York—at the time (1858) the largest stone barn in North America (196 feet by 50 feet, and 5 stories high)—did not seem out of scale in its environmental context. The Shakers' adherence to the teachings of their Millenial Laws, their special communal needs, and their physical separation from the rest of society seem to have unshackled them from the constraints of contemporary design. Perhaps their finest achievement in the construction of farm buildings was their round stone barn at Hancock, Massachusetts. It is a study in marvelous simplicity, and is, itself, a departure from the usual rectangular Shaker barn. It was thought to be a design directly attributable to God and therefore could not be duplicated by any other Shaker community. Its superior organization and efficiency anticipated scientific dairying by several decades.

Many secular builders tried their own interpretations of the round barn. Smaller wooden models cropped up in Pennsylvania, Ontario, the Midwest (it was given a domed roof in Michigan's Lower Peninsula), and California. Although they were practical, labor-saving farm improvements, round barns were expensive and usually required specialized knowledge to build. The advantages of the circular form were available for less cost in the octagonal design which, because of its straight walls, was less of a challenge to carpenters.

Fascination with the octagon reached a high point in the 1850s following publication of *A Home for All* by phrenologist and health-writer Orson Squire Fowler. The octagon was the best design for a house, claimed Fowler, because it contained one-fifth more floor space than a square house with walls of the same length. What's more, it was more beautiful than a square dwelling because its geometry approached that of the sphere, Nature's governing form.

Many formal and stylistic changes in barn construction were the result of a growing interest by landscape architects and designers in the rural environment. A. J. Downing, for example, presented the most influential case for integrating geography and architecture in a tasteful manner; for him the appropriate style was expressed in the Gothic mode. His widely-read book, *The Architecture of Country Houses* (1850), and those of several other followers initiated a concern among farmers and rural builders for aesthetic considerations in new and remodeled barns. As farmers responded to the growing trend toward specialization and the appeal of scientific techniques during the blossoming Victorian era, their barns were often rebuilt in the predominant contemporary fashion. Simple gable

ends became mazes of vergeboard topped by finials; great doors were transformed into cathedral arches. Healthful ventilation, the desideratum of modern farming, was accomplished by capping air shafts with vents or cupolas, some of which became absurdly ornate. A few of the more soaring and ambitious barns were erected by gentlemen farmers for their own amusement or for experimental farming.

Domestic architecture in the mid to late 19th century did much more than alter a barn's stylistic manifestations. Truss beams, long known to bridge builders and commonly used in balloon framing to support the relatively small roofs of houses, replaced the great cross beams and posts heretofore needed to support huge barn roofs. This not only permitted the building of barns with vaulting gambrels of enormous capacity, but allowed clear space for hay-handling equipment in the loft. Unfortunately, gambrel roof construction required large amounts of labor and was therefore expensive. More recently, a cheaper prefabricated Gothic roof (called "Already Cut" and sold by Sears in the 1920s), combined with pre-made walls, replaced the gambrel. This in turn has been largely replaced by all metal, rigid-arch buildings of only one level. The transformation of the rural landscape wrought by these late 19th- and early 20th-century developments included another equally-important change: the introduction of the silo.

The silo did not exist before the 1870s, at least not in the form of the familiar, stout cylinder we picture appended to a barn. In its place was a covered pit, sometimes within the barn, designed to keep cattle fodder succulent and nutritious. The need for an air-tight container led first to stone, wood, and then concrete silos before the First World War. Today's efficient blue towers are like giant steel-jacketed thermos bottles.

Despite the economic pressures to renovate or replace old barns with new, many old ones still carry their weight and remain vital to the operation of a farm, be it in Tennessee or Alberta. Their basic functions have not changed radically in more than three centuries, and chances are, barring the inexorable march of suburbia, they will not change all that much in the coming years.

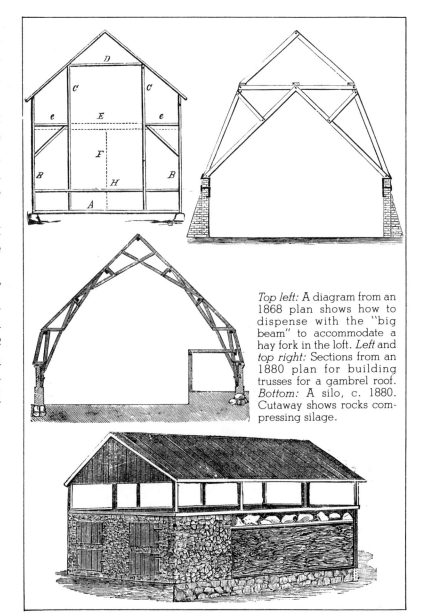

Top left: A diagram from an 1868 plan shows how to dispense with the "big beam" to accommodate a hay fork in the loft. *Left* and *top right:* Sections from an 1880 plan for building trusses for a gambrel roof. *Bottom:* A silo, c. 1880. Cutaway shows rocks compressing silage.

17th-century primitive barn, Burlington County, New Jersey. Few log structures of such vintage remain standing in North America. This illustration was drawn in the 1930s. Further details can be seen on pages 18-19.

primitive barns

The primitive barn combines all that is quintessentially rustic: crudely-notched and rough-hewn logs, hand-wrought hardware (if indeed there was any), and coarse shingles, all assembled in the simplest manner. Its character has long been perceived as hardy, spare, and virtuous in its unity with its rugged surroundings. The log barn possesses such enviable qualities that many 19th century politicians claimed a special blessing for having been born in at least its domestic equivalent.

Most of North America's first barns were built of logs. The logs were usually notched to form secure corners or, in the case of French Canadian barns of the 17th century, simply beveled at the ends and held in place by a corner post driven into the ground and grooved to receive the next log. These timbers required little preparation beyond de-barking and squaring, although neither operation was necessary. The manner of notching depended upon ethnic tradition and on one's skill with an axe. A carefully-cut dovetail not only made a very solid corner, but the bevel drained moisture away from the interior and helped protect the logs from wet rot. Some builders, especially the 17th-century Swedes, butted logs so tightly that chinking was hardly necessary. Usually daub or straw was stuffed into the cracks. If hay were stored in a loft above the livestock, the spaces were left open on that level to permit ventilation. The roof was made of sod, thatch, bark, hand-riven shingles, or even lengths of hollowed log sections laid between the top course of logs and the ridge.

Log barns had several advantages over frame construction. They could be erected relatively quickly by one man, if necessary. They were warmer (if tightly chinked), less flammable, and lasted much longer without having to be painted or otherwise maintained. On the other hand, they lacked the flexibility of framing which allowed for a building of almost any size and shape.

The following plans illustrate primitive barns from nearly every region of the United States. Some are typical of those found on the Canadian frontiers. Although the earliest example dates back to the 1600s and the latest was built in the mid-1800s, they share a common point: each was built by a pioneer who was among the first to settle each region.

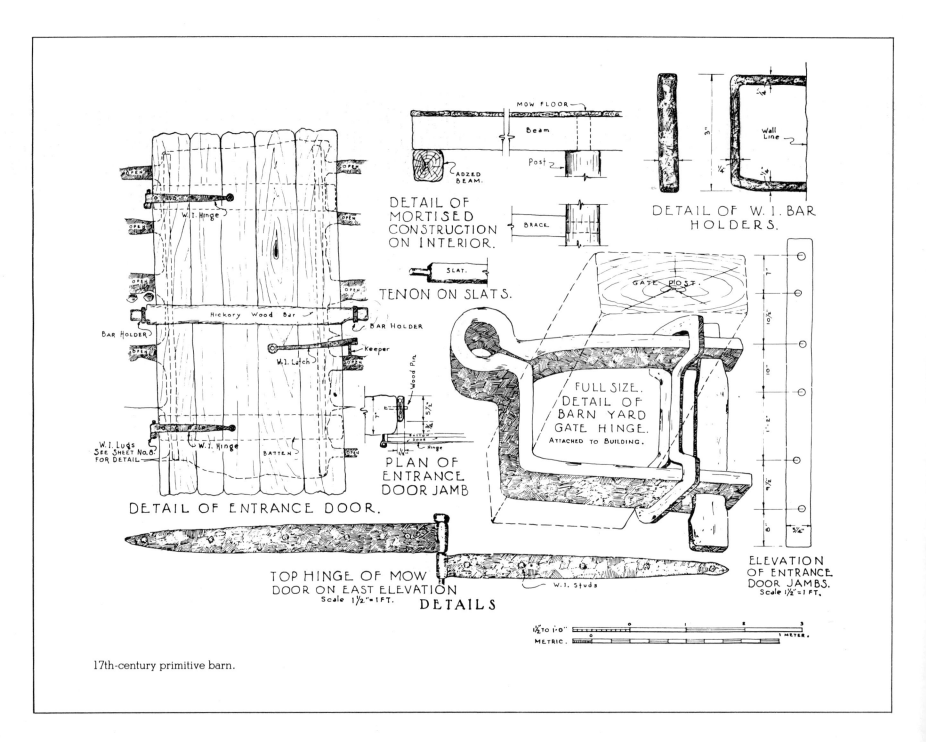

DETAIL OF MORTISED CONSTRUCTION ON INTERIOR.

MOW FLOOR

Beam

Post

ADZED BEAM.

BRACE.

SLAT.

TENON ON SLATS.

DETAIL OF W.I. BAR HOLDERS.

Wall Line

3"

W.I. Hinge

OPEN

Bar Holder

Hickory Wood Bar

Bar Holder

Keeper

W.I. Latch

W.I. Lugs SEE SHEET NO. 8. FOR DETAIL

W.I. Hinge

BATTEN

DETAIL OF ENTRANCE DOOR.

Wood Pin

Batten Door Hinge

PLAN OF ENTRANCE DOOR JAMB

GATE POST.

FULL SIZE. DETAIL OF BARN YARD GATE HINGE. Attached to Building.

ELEVATION OF ENTRANCE DOOR JAMBS. Scale 1½"=1 FT.

TOP HINGE OF MOW DOOR ON EAST ELEVATION Scale 1½"=1 FT.

W.I. Studs

DETAILS

1½ TO 1·0"

METRIC.

1 METER.

17th-century primitive barn.

18

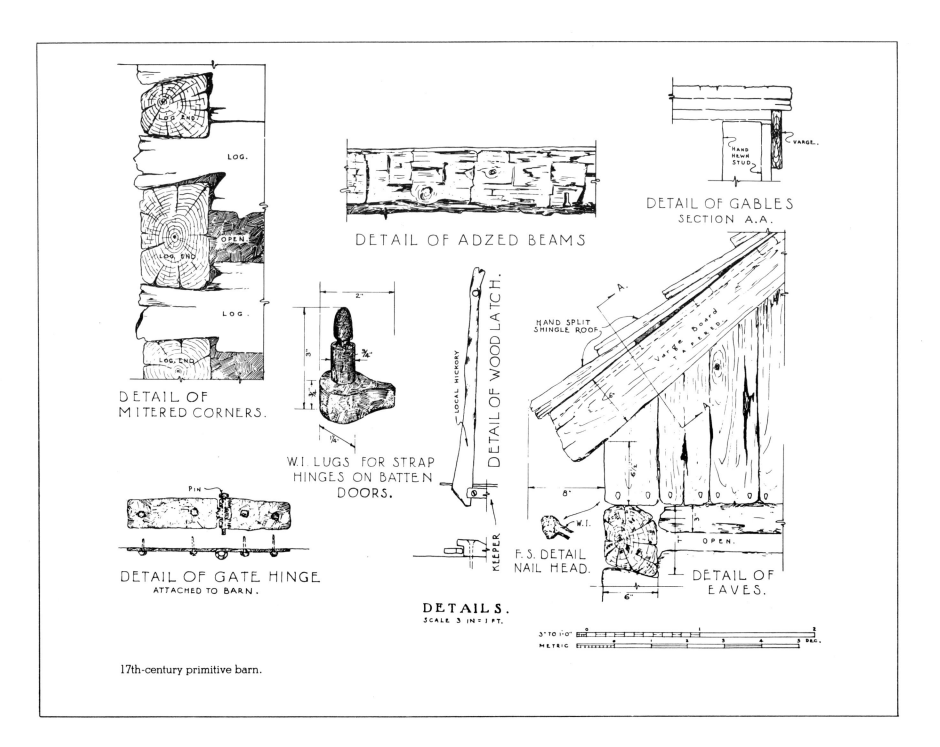

DETAIL OF MITERED CORNERS.

LOG.

LOG.

OPEN.

LOG. END

LOG.

LOG. END

DETAIL OF ADZED BEAMS

DETAIL OF GABLES
SECTION A.A.

VARGE.

HAND HEWN STUD

W.I. LUGS FOR STRAP HINGES ON BATTEN DOORS.

2"

3"

¾"

¾"

¼"

LOCAL HICKORY

DETAIL OF WOOD LATCH.

KEEPER

HAND SPLIT SHINGLE ROOF

A.

Varge Board TAPERED.

6"

8"

6½"

3"

OPEN.

6"

DETAIL OF EAVES.

PIN

DETAIL OF GATE HINGE
ATTACHED TO BARN.

W.I.

F. S. DETAIL
NAIL HEAD.

DETAILS.
SCALE 3 IN.= 1 FT.

3" TO 1'-0"   0   1   2
METRIC   1   2   3   4   5 DEC.

17th-century primitive barn.

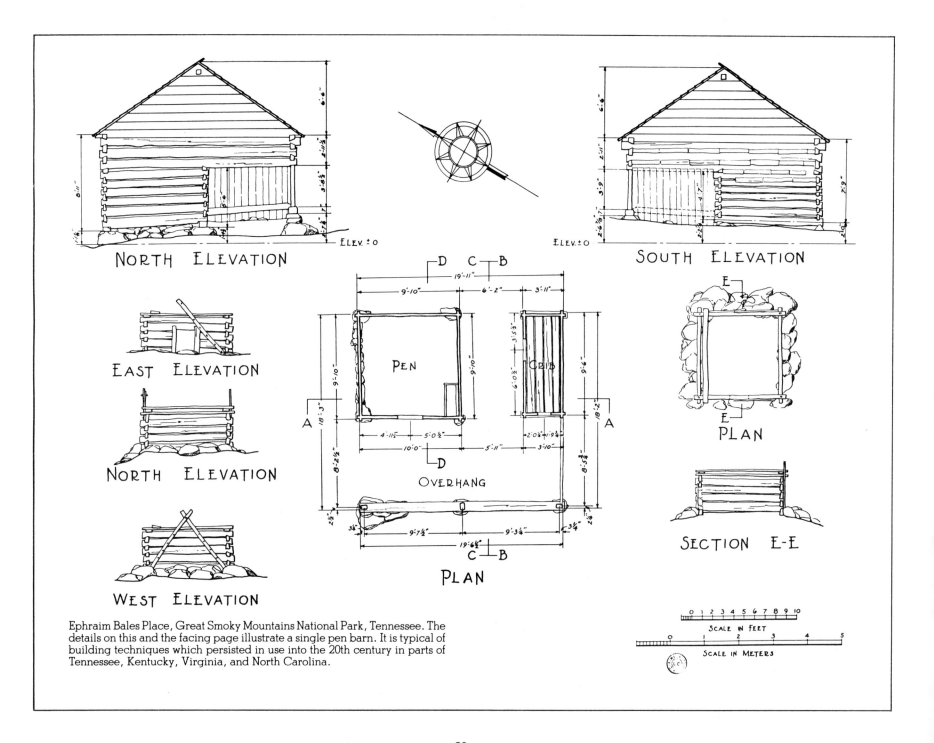

NORTH ELEVATION

SOUTH ELEVATION

EAST ELEVATION

NORTH ELEVATION

WEST ELEVATION

PEN

CRIB

OVERHANG

PLAN

E PLAN

SECTION E-E

Ephraim Bales Place, Great Smoky Mountains National Park, Tennessee. The details on this and the facing page illustrate a single pen barn. It is typical of building techniques which persisted in use into the 20th century in parts of Tennessee, Kentucky, Virginia, and North Carolina.

0 1 2 3 4 5 6 7 8 9 10
SCALE IN FEET

0 1 2 3 4 5
SCALE IN METERS

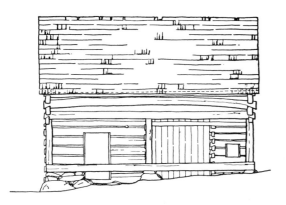

WEST ELEVATION

SECTION D-D

EAST ELEVATION

SECTION A-A

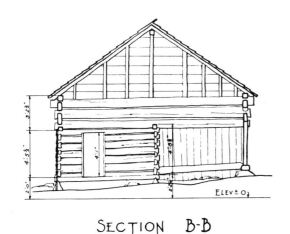

SECTION B-B

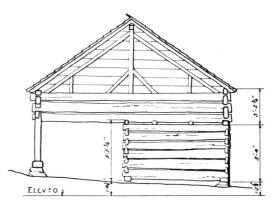

SECTION C-C

NOTE

PEN WALLS - SPLIT CHESTNUT LOGS 3" TO 4" THICK.
CRIB WALLS - HEWN CHESTNUT LOGS 5" THICK.
THREE UPPER ROUNDS OF BARN - CHESTNUT HEWN
4" TO 7½" THICK
CRIB FLOOR - CHESTNUT PUNCHEONS.
RAFTERS - CHESTNUT POLES 4" IN DIAMETER.
VERTICAL SHEATHING OVER OPENING BETWEEN
PEN AND CRIB ON WEST ELEVATION, & SOUTH
END OF OVER-HANG & SAWN POPLAR BOARDS
SHEATHING BOTH GABLES WERE REMOVED —
WHEN THE PROPERTY WAS VACATED
FEED TROUGH - SPLIT LOG. HOLLOWED OUT.
ROOF - RIVED OAK BOARDS LAID AS SHINGLES.

0 1 2 3 4 5 6 7 8 9 10
SCALE IN FEET

0 1 2 3 4 5
SCALE IN METERS

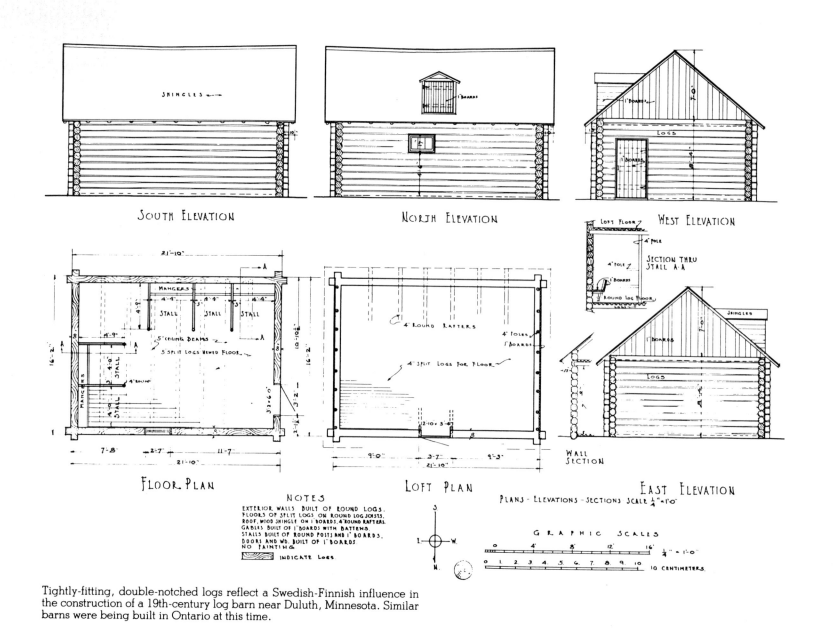

SOUTH ELEVATION

NORTH ELEVATION

WEST ELEVATION

SECTION THRU STALL A-A

FLOOR PLAN

LOFT PLAN

EAST ELEVATION

WALL SECTION

NOTES

EXTERIOR WALLS BUILT OF ROUND LOGS.
FLOORS OF SPLIT LOGS ON ROUND LOG JOISTS.
ROOF, WOOD SHINGLE ON 1" BOARDS, 4" ROUND RAFTERS.
GABLES BUILT OF 1" BOARDS WITH BATTENS
STALLS BUILT OF ROUND POSTS AND 1" BOARDS.
DOORS AND WD. BUILT OF 1" BOARDS
NO PAINTING

INDICATE LOGS

PLANS - ELEVATIONS - SECTIONS SCALE ¼"=1'-0"

GRAPHIC SCALES

Tightly-fitting, double-notched logs reflect a Swedish-Finnish influence in the construction of a 19th-century log barn near Duluth, Minnesota. Similar barns were being built in Ontario at this time.

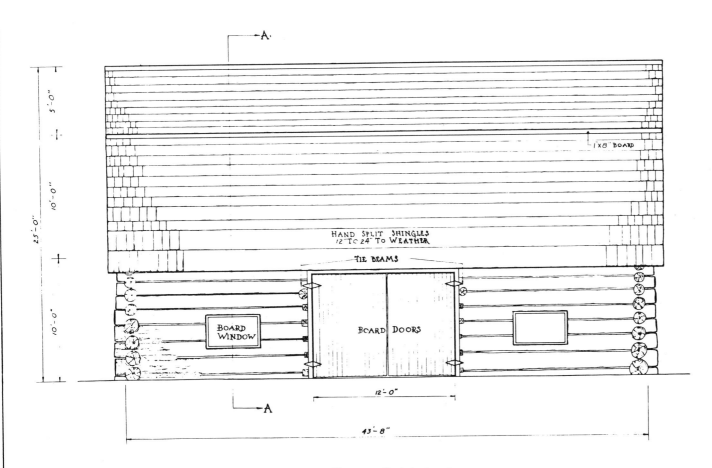

A

5'-0"

10'-0"

25'-0"

1 x 8" BOARD

HAND SPLIT SHINGLES
12" TO 24" TO WEATHER

TIE BEAMS

10'-0"

BOARD
WINDOW

BOARD DOORS

A

12'-0"

43'-8"

NORTH ELEVATION
SCALE ¼" = 1'-0"

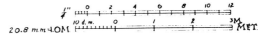

Old log barn, Potter County, Pennsylvania. The gambrel roof was more commonly found on houses than on barns in the late 18th century. The appearance of stability is enhanced by placement of larger-diameter logs at the bottom of walls.

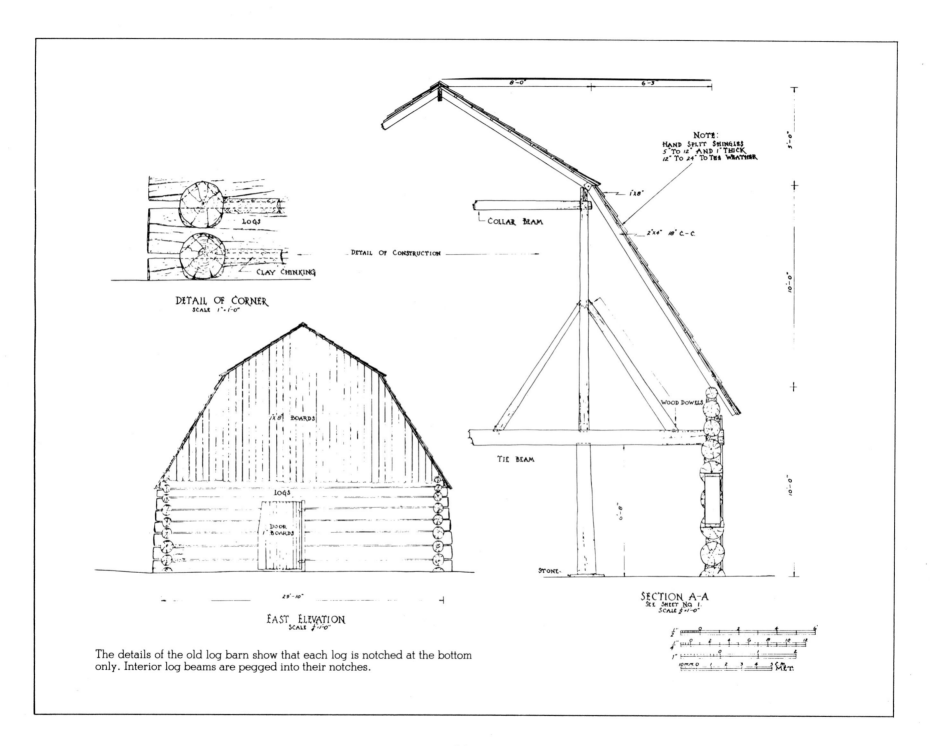

LOGS

CLAY CHINKING

DETAIL OF CORNER
SCALE 1"=1'-0"

DETAIL OF CONSTRUCTION

NOTE:
HAND SPLIT SHINGLES
5" TO 12" AND 1" THICK
12" TO 24" TO THE WEATHER

COLLAR BEAM

1"x8"

2"x4" 18" C-C.

8'-0"   6'-3"

5'-0"

10'-0"

WOOD DOWELS

TIE BEAM

10'-0"

5'-0"

STONE

SECTION A-A
SEE SHEET NO. 1.
SCALE 1/2"=1'-0"

1'x8" BOARDS

LOGS

DOOR
1" BOARDS

29'-10"

EAST ELEVATION
SCALE 1/4"=1'-0"

The details of the old log barn show that each log is notched at the bottom
only. Interior log beams are pegged into their notches.

24

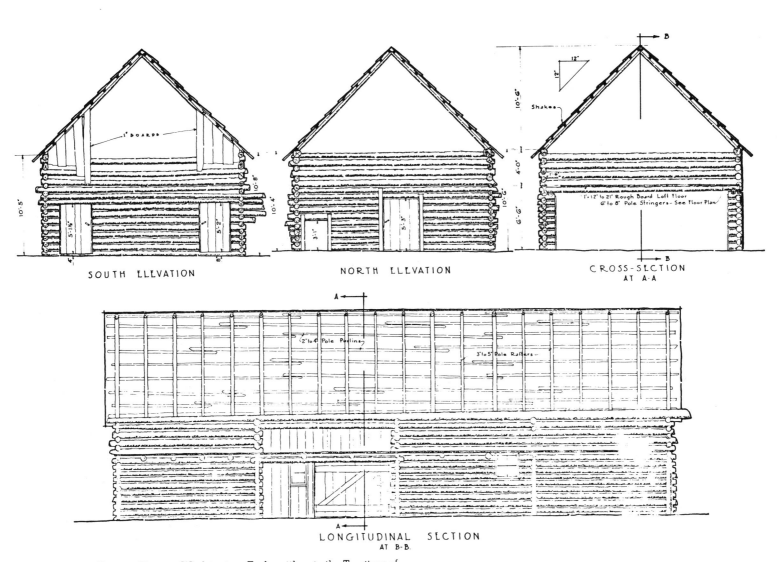

SOUTH ELEVATION

NORTH ELEVATION

CROSS-SECTION
AT A-A

LONGITUDINAL SECTION
AT B-B.

Clugston barn, Stevens County, Washington. Early settlers in the Territory of Washington built primitive barns in the mid-1800s, but the gold rush in the 1860s and the opening of rail travel to the East in 1884 rapidly transformed both the architecture and the economy of the region.

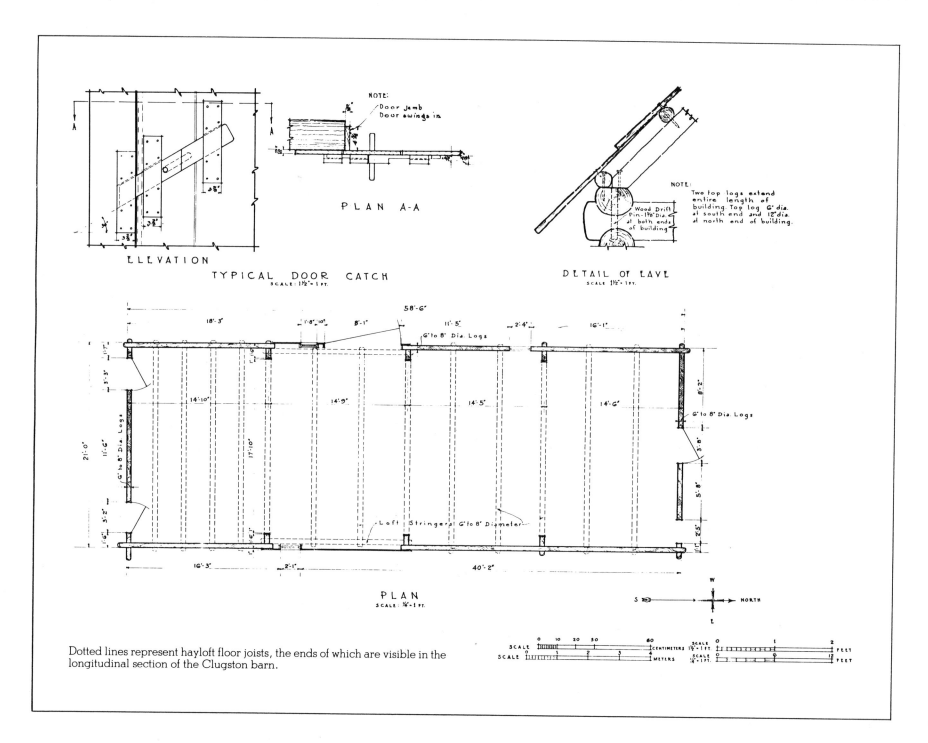

NOTE:
Door Jamb
Door swings in

PLAN A-A

NOTE:
Two top logs extend
entire length of
building. Top log 6" dia.
at south end and 12" dia.
at north end of building.

Wood Drift
Pin 1⅜" Dia.
at both ends
of building

ELEVATION

TYPICAL DOOR CATCH
SCALE: 1½" = 1 FT.

DETAIL OF EAVE
SCALE 1½" = 1 FT.

58'-6"

18'-3"   1'-8" 10"   8'-1"   11'-5"   2'-4"   16'-1"

6" to 8" Dia. Logs

3'-3"

21'-0"   11'-6"

6" to 8" Dia. Logs

14'-10"   17'-10"   14'-9"   14'-5"   14'-6"

8'-2"

6" to 8" Dia. Logs

3'-8"

5'-8"

Loft Stringers 6" to 8" Diameter

3'-2"   2'-5"

16'-3"   2'-1"   40'-2"

PLAN
SCALE: ¼" = 1 FT.

W
S ⊙→   NORTH
E

Dotted lines represent hayloft floor joists, the ends of which are visible in the
longitudinal section of the Clugston barn.

SCALE   0 10 20 30   60   SCALE 0   1   2
SCALE   CENTIMETERS 1½" = 1 FT.   FEET
SCALE   0   1   2   3   4   METERS   SCALE 0   6   12
                                              ¼" = 1 FT.   FEET

26

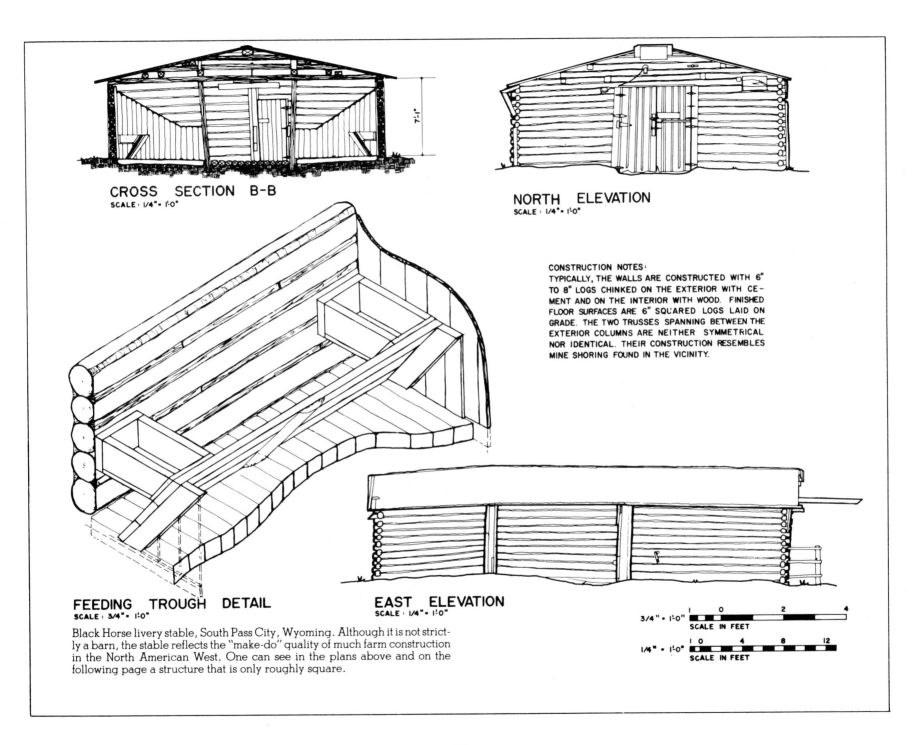

CROSS SECTION B-B
SCALE : 1/4" = 1'-0"

NORTH ELEVATION
SCALE : 1/4" = 1'-0"

7'-1"

CONSTRUCTION NOTES:
TYPICALLY, THE WALLS ARE CONSTRUCTED WITH 6"
TO 8" LOGS CHINKED ON THE EXTERIOR WITH CE-
MENT AND ON THE INTERIOR WITH WOOD. FINISHED
FLOOR SURFACES ARE 6" SQUARED LOGS LAID ON
GRADE. THE TWO TRUSSES SPANNING BETWEEN THE
EXTERIOR COLUMNS ARE NEITHER SYMMETRICAL
NOR IDENTICAL. THEIR CONSTRUCTION RESEMBLES
MINE SHORING FOUND IN THE VICINITY.

FEEDING TROUGH DETAIL
SCALE : 3/4" = 1'-0"

EAST ELEVATION
SCALE : 1/4" = 1'-0"

3/4" = 1'-0"   1 0   2   4
SCALE IN FEET

1/4" = 1'-0"   1 0   4   8   12
SCALE IN FEET

Black Horse livery stable, South Pass City, Wyoming. Although it is not strictly a barn, the stable reflects the "make-do" quality of much farm construction
in the North American West. One can see in the plans above and on the
following page a structure that is only roughly square.

27

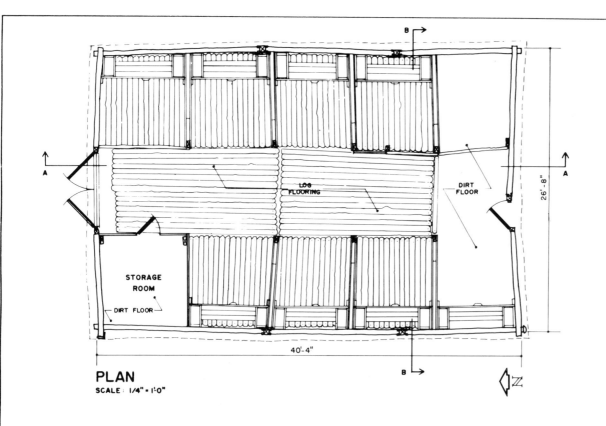

26'-8"

LOG FLOORING

DIRT FLOOR

STORAGE ROOM

DIRT FLOOR

40'-4"

PLAN
SCALE: 1/4" = 1'-0"

N

THE BLACK HORSE LIVERY STABLE, BUILT c.1868 FOR GEORGE DUNN, IS THE ONLY REMAINING ONE OF THREE LIVERIES. AFTER 1893, THE STRUCTURE WAS SOLD AS PART OF THE WOLVERINE LODE CLAIM AND WAS DEEDED TO A SERIES OF OWNERS UNTIL 1901 WHEN IT WAS PURCHASED BY THE SHERLOCK FAMILY WHO SOLD IT TO FRED STRATTON IN 1948. JOHN WOODRING, WHO ACQUIRED THE PROPERTY IN 1955, SOLD IT IN 1966 TO THE WYOMING 75th ANNIVERSARY COMMISSION, INC. IN 1967 IT BECAME PART OF THE OLD SOUTH PASS CITY HISTORICAL PRESERVE ESTABLISHED BY THE WYOMING LEGISLATURE. AT THE TIME OF THE PURCHASE BY THE COMMISSION, A LARGE SHED BUILT c.1900 ADJACENT TO THE WEST AND SOUTH WALLS WAS RAZED.

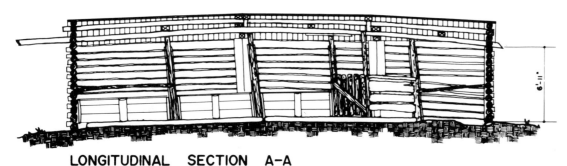

6'-11"

LONGITUDINAL SECTION A-A
SCALE: 1/4" = 1'-0"

Black Horse livery stable.

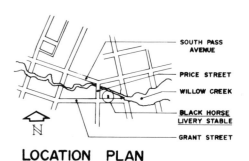

SOUTH PASS AVENUE

PRICE STREET

WILLOW CREEK

BLACK HORSE LIVERY STABLE

GRANT STREET

LOCATION PLAN
SCALE: 1" = 500'

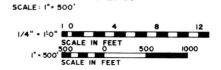

1/4" = 1'-0"   1 0   4   8   12
SCALE IN FEET
1" = 500'   500   0   500   1000
SCALE IN FEET

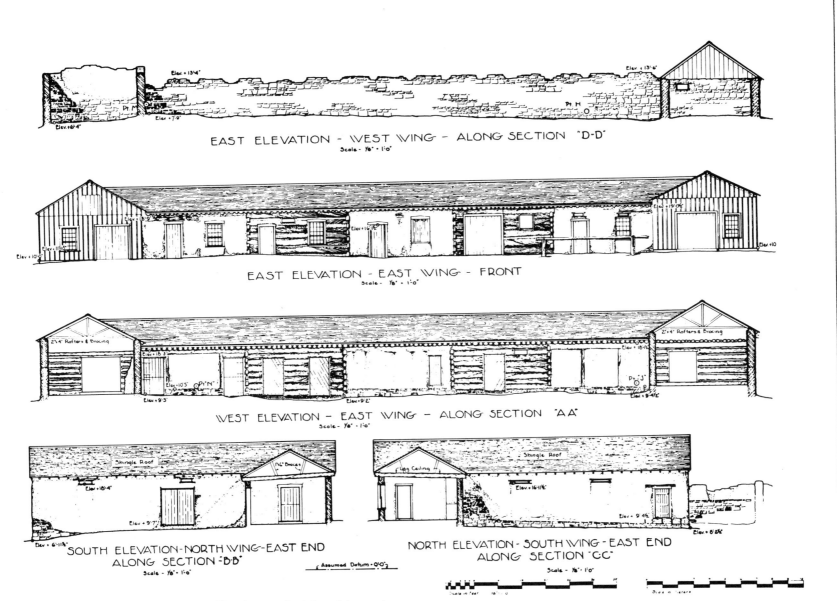

EAST ELEVATION - WEST WING - ALONG SECTION "D-D"
Scale - 1/8" = 1'-0"

EAST ELEVATION - EAST WING - FRONT
Scale - 1/8" = 1'-0"

WEST ELEVATION - EAST WING - ALONG SECTION "A-A"
Scale - 1/8" = 1'-0"

SOUTH ELEVATION-NORTH WING-EAST END
ALONG SECTION "B-B"
Scale - 1/8" = 1'-0"

NORTH ELEVATION-SOUTH WING-EAST END
ALONG SECTION "C-C"
Scale - 1/8" = 1'-0"

Tipton barn, Tiptonville, New Mexico. The elongated patchwork is a curious mix of log, board and batten, adobe, sheet iron, and stone construction.

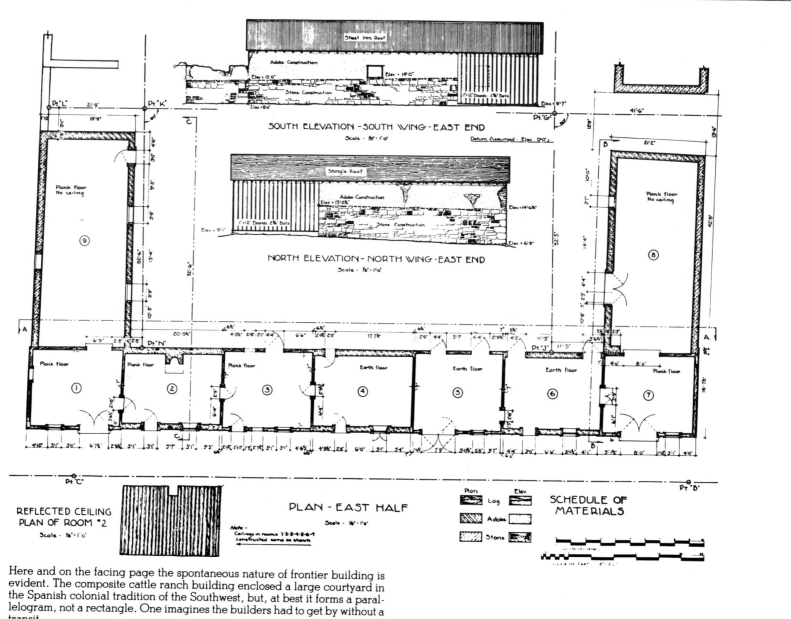

SOUTH ELEVATION - SOUTH WING - EAST END
Scale · ⅛"·1'0"   Datum Assumed - Elev. 0'0"

NORTH ELEVATION - NORTH WING - EAST END
Scale · ⅛"·1'0"

REFLECTED CEILING
PLAN OF ROOM #2
Scale · ⅛"·1'0"

PLAN - EAST HALF
Scale · ⅛"·1'0"

Note -
Ceilings in rooms 1-2-3-4-5-6-7
constructed same as shown

SCHEDULE OF
MATERIALS

	Plan	Elev.
Log		
Adobe		
Stone		

Here and on the facing page the spontaneous nature of frontier building is
evident. The composite cattle ranch building enclosed a large courtyard in
the Spanish colonial tradition of the Southwest, but, at best it forms a paral-
lelogram, not a rectangle. One imagines the builders had to get by without a
transit.

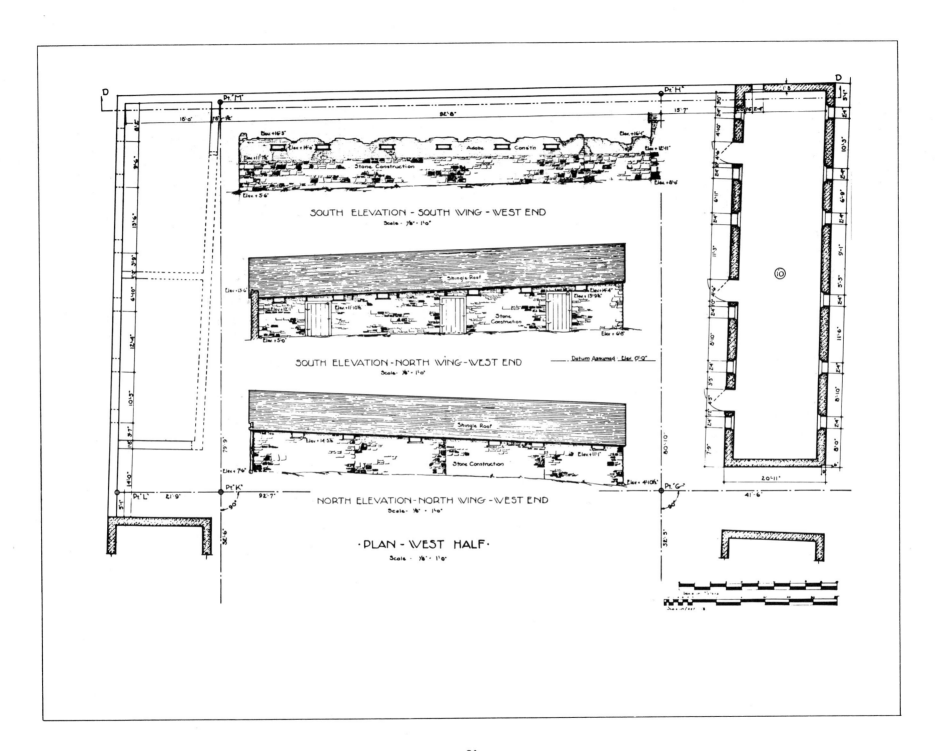

SOUTH ELEVATION - SOUTH WING - WEST END
Scale- ⅛" = 1'0"

SOUTH ELEVATION - NORTH WING - WEST END
Scale- ⅛" = 1'0"

_____ Datum Assumed - Elev 0'-0"

NORTH ELEVATION - NORTH WING - WEST END
Scale- ⅛" = 1'0"

·PLAN - WEST HALF·
Scale - ⅛" = 1'0"

31

· BIRD'S · EYE · VIEW · OF · FARM · GROUP ·

The John Zabriskie farm group, Paramus, New Jersey. Details of the barn visible at the upper left are on pages 50-51.

early english and dutch barns

Some early frame barns appear to be nearly as crude as their log predecessors, whereas others built in New England as early as the 1650s were clearly years ahead of their contemporaries. Regardless of the apparent level of sophistication, frame barns were most certainly of a different order than log structures. Between the felling of a tree and the nailing down of the final shingle were many intermediate steps not required in log construction. Not only did a log have to be squared to a consistent dimension, but it had to be cut to a specific length and, if it were a post or beam, mortised or otherwise prepared to attach correctly to another similarly-cut piece. The siding, be it weatherboard, board and batten, or tongue and groove, usually had to be cut by hand. (Sawmills did not produce substantial quantities of lumber until water power was replaced by steam in the 1830s.) The individual framing members, or bents, had to be test-assembled on the ground and labeled for later, final assembly. All this required the efforts of many skilled hands and the expenditure of a great deal of time. Despite the fabled intensity of one-day "barn raisings," most of the early barns were not the result of great community efforts.

Although many early frame barns shared the tripartite "English" configuration (the doors opened onto a threshing floor flanked by two mows or stock aisles parallel to the gable ends), there were also a number of variations to which the following pages attest. The English barn, modified at times, remained popular well into the late 19th century when it was built on a much larger scale in the Midwest and on the Canadian plains.

Dutch barns of the 17th century were not as widespread or as popular as the English, but are no less structurally interesting. Through the gable-end doors one can see the massive anchor beams supporting the hayloft. In cross section they can measure up to 12 by 24 inches and 30 feet long and, depending on their number, they divide the barn into anywhere from three to seven bays parallel to the gable ends. The roof lines of Dutch barns are of two slopes (not the gambrels now associated with Dutch-Colonial houses) and are tall enough to accommodate a large amount of hay. The hay is loaded from the ground level and piled on movable poles laid across the anchor beams. It is not unusual to find that some of these poles are the same spike-tipped pikes used to help pivot the assembled bent into position during construction.

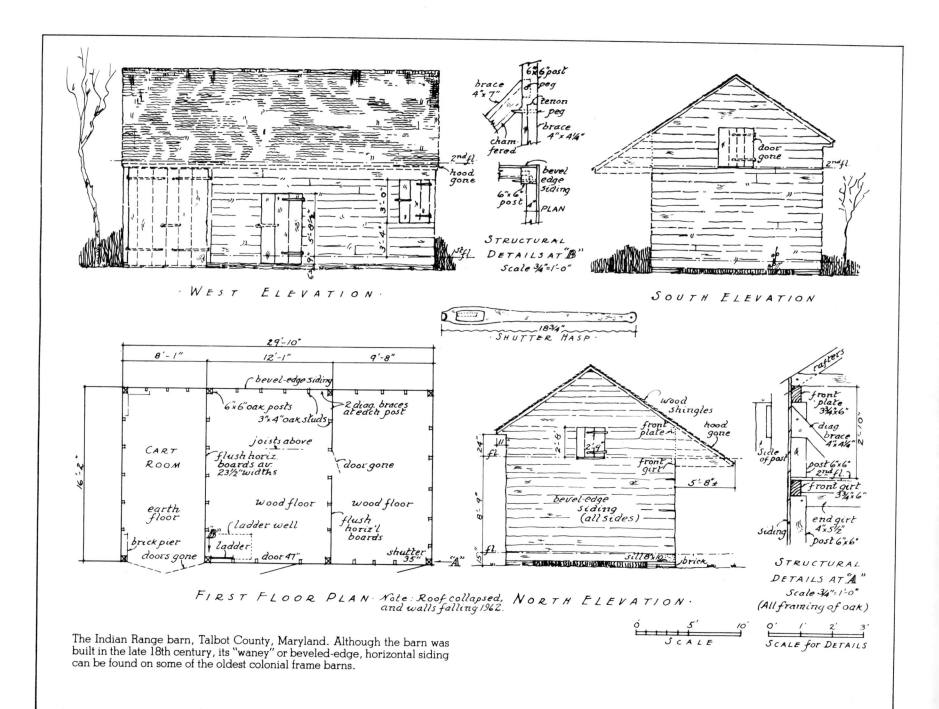

· W E S T   E L E V A T I O N ·

2nd fl.
hood
gone
1st fl.

brace
4"x 7"
cham-
fered
6"x6" post
peg
tenon
peg
brace
4"x 4¼"

bevel
edge
siding
6"x 6"
post   PLAN

STRUCTURAL
DETAILS AT "B"
Scale ¾"=1'-0"

door
gone

2nd fl.

S O U T H   E L E V A T I O N

18¾"
· SHUTTER HASP ·

FIRST FLOOR PLAN

29'-10"
8'-1"   12'-1"   9'-8"
16'-2"

bevel-edge siding
6"x6" oak posts
3"x 4" oak studs
2 diag. braces
at each post
joists above
flush horiz.
boards av.
23½" widths
door gone

CART
ROOM

earth
floor

brick pier
doors gone

"B"
ladder well
ladder
door 47"

wood floor

wood floor
flush
horiz'l
boards
shutter
35"

"A"

Note: Roof collapsed,
and walls falling 1962.

N O R T H   E L E V A T I O N ·

wood
shingles
hood
gone
front
plate
front
girt
5'-8"±

2nd
fl.
11"
24
2'-6"
2'9"

8'-9"

1st
fl.

bevel-edge
siding
(all sides)

sill 8"x10"   brick

rafters

front
plate
3¼"x 6"
diag
brace
4"x4¼"
side
of post
post 6"x6"
2nd fl.
front girt
3¼"x 6"
end girt
4"x 5½"
post 6"x6"
siding

2'-0"

STRUCTURAL
DETAILS AT "A"
Scale ¾"=1'-0"
(All framing of oak)

0   5'   10'
SCALE

0'   1'   2'   3'
SCALE for DETAILS

The Indian Range barn, Talbot County, Maryland. Although the barn was
built in the late 18th century, its "waney" or beveled-edge, horizontal siding
can be found on some of the oldest colonial frame barns.

34

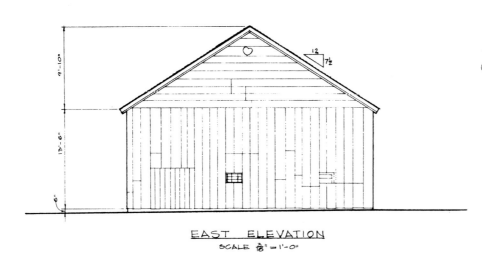

EAST ELEVATION
SCALE $\frac{3}{16}$"=1'-0"

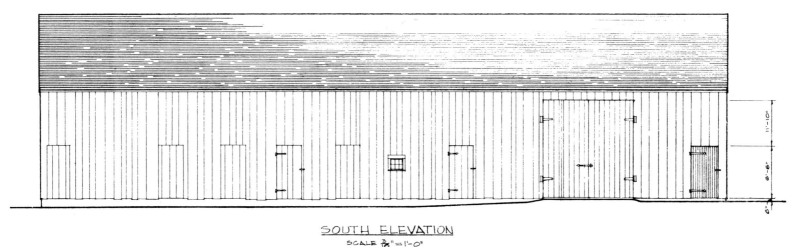

SOUTH ELEVATION
SCALE $\frac{3}{16}$"=1'-0"

The Hunt-Hosmer barn, Concord, Massachusetts. Patriots passed by this barn on their way to engage British troops in April, 1775. The siding is vertical tongue and groove; wood shingles cover the roof.

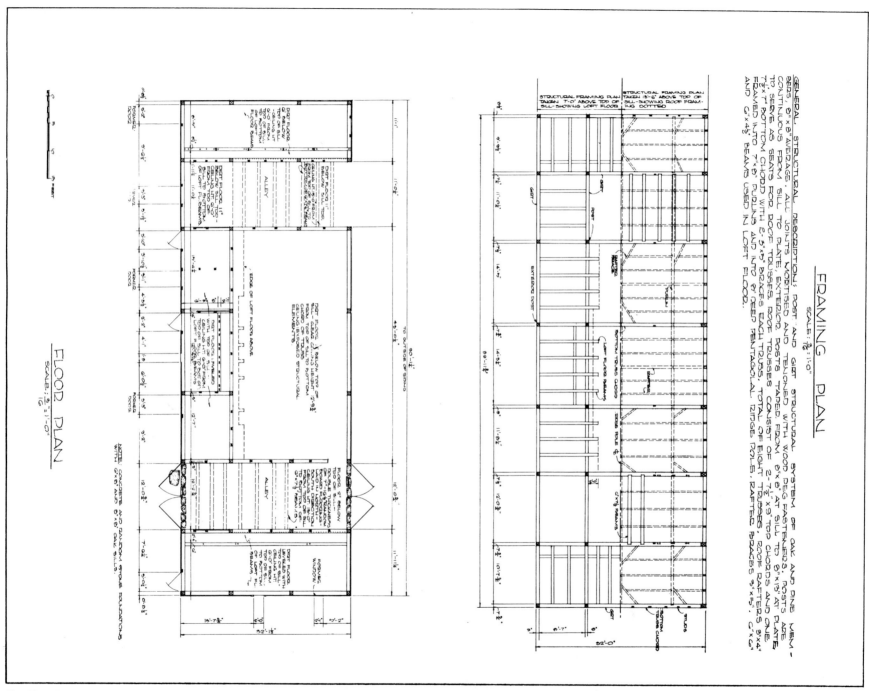

The Hunt-Hosmer barn.

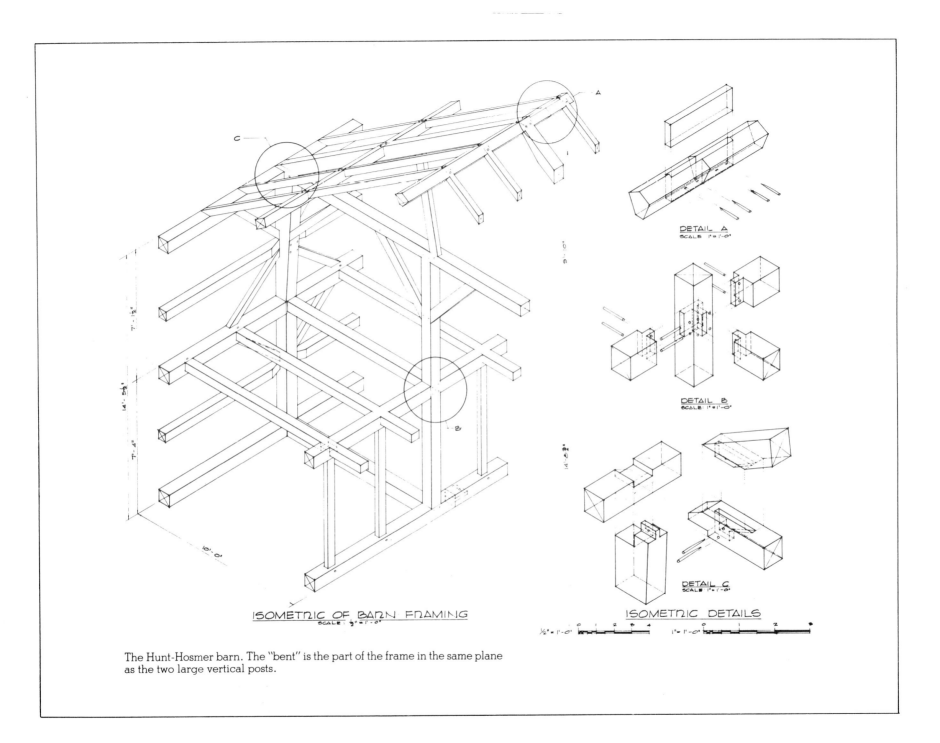

ISOMETRIC OF BARN FRAMING
SCALE: ½" = 1'-0"

DETAIL A
SCALE: 1" = 1'-0"

DETAIL B
SCALE: 1" = 1'-0"

DETAIL C
SCALE: 1" = 1'-0"

ISOMETRIC DETAILS

½" = 1'-0"   1" = 1'-0"

The Hunt-Hosmer barn. The "bent" is the part of the frame in the same plane as the two large vertical posts.

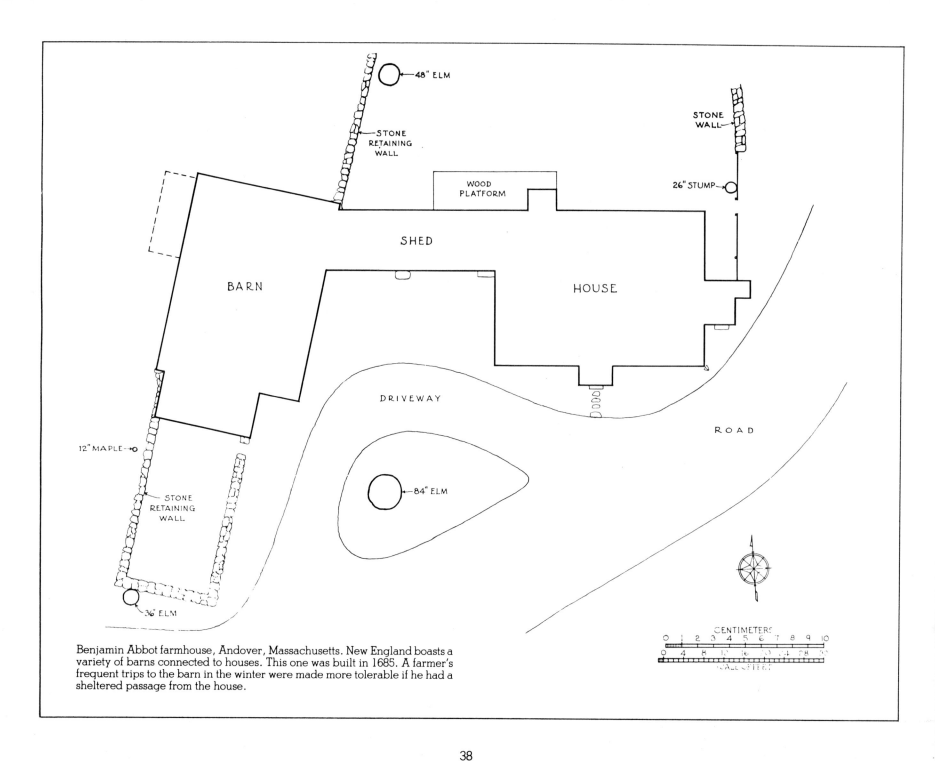

48" ELM

STONE
RETAINING
WALL

STONE
WALL

WOOD
PLATFORM

26" STUMP

SHED

BARN

HOUSE

DRIVEWAY

ROAD

12" MAPLE

STONE
RETAINING
WALL

84" ELM

36" ELM

CENTIMETERS
0 1 2 3 4 5 6 7 8 9 10

0   4   8   12   16   20   24   28   32
SCALE, FEET

Benjamin Abbot farmhouse, Andover, Massachusetts. New England boasts a
variety of barns connected to houses. This one was built in 1685. A farmer's
frequent trips to the barn in the winter were made more tolerable if he had a
sheltered passage from the house.

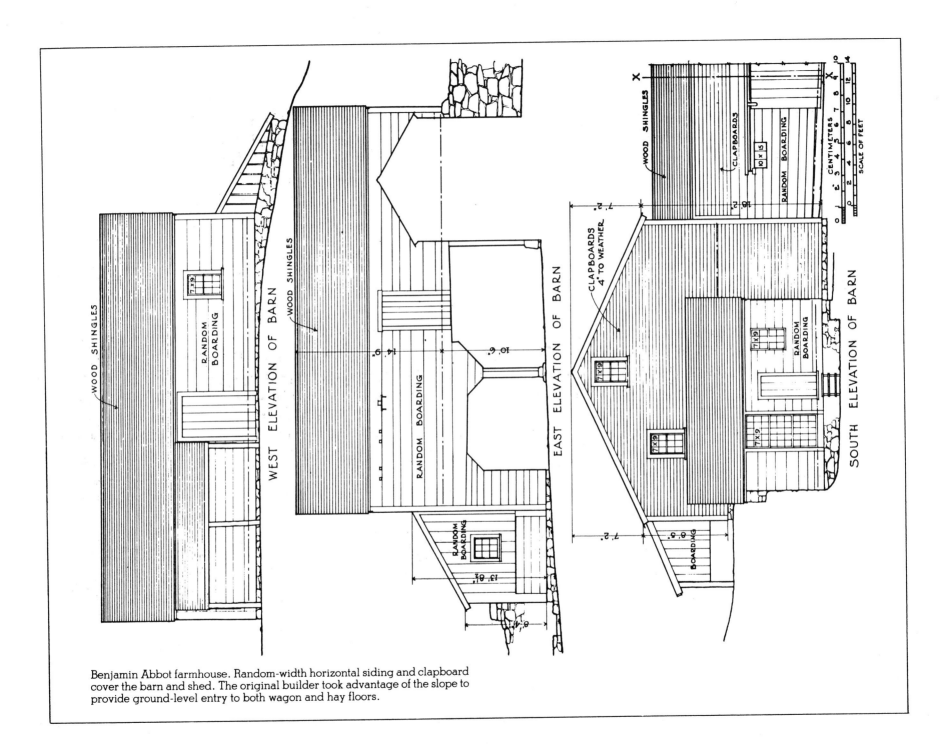

Benjamin Abbot farmhouse. Random-width horizontal siding and clapboard cover the barn and shed. The original builder took advantage of the slope to provide ground-level entry to both wagon and hay floors.

39

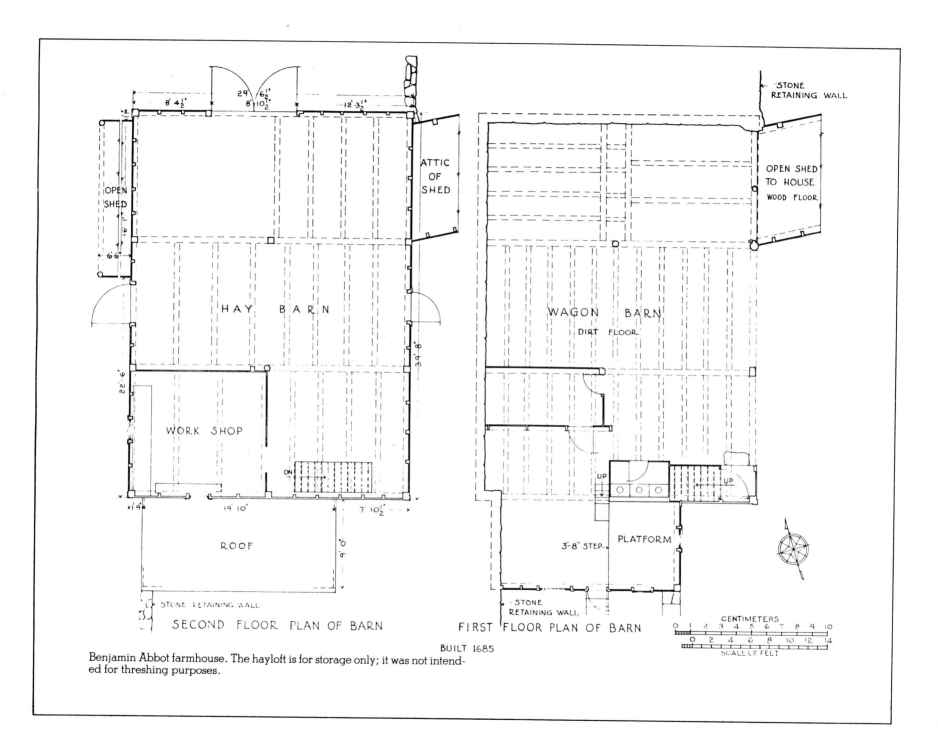

29' 6½"
8' 4½" 8' 10½" 12' 3½"

STONE RETAINING WALL

OPEN SHED

ATTIC OF SHED

OPEN SHED TO HOUSE
WOOD FLOOR

H A Y   B A R R N

WAGON   BARN
DIRT FLOOR

6' 1"

6' 6"

34' 8"

22' 6"

WORK SHOP

UP

DN

UP

1' 9"
19' 10"
7' 10½"

9' 0"

3'-8" STEP

PLATFORM

ROOF

STONE RETAINING WALL

STONE RETAINING WALL

CENTIMETERS
0 1 2 3 4 5 6 7 8 9 10
0 2 4 6 8 10 12 14
SCALE OF FEET

SECOND FLOOR PLAN OF BARN

FIRST FLOOR PLAN OF BARN

BUILT 1685

Benjamin Abbot farmhouse. The hayloft is for storage only; it was not intend-
ed for threshing purposes.

40

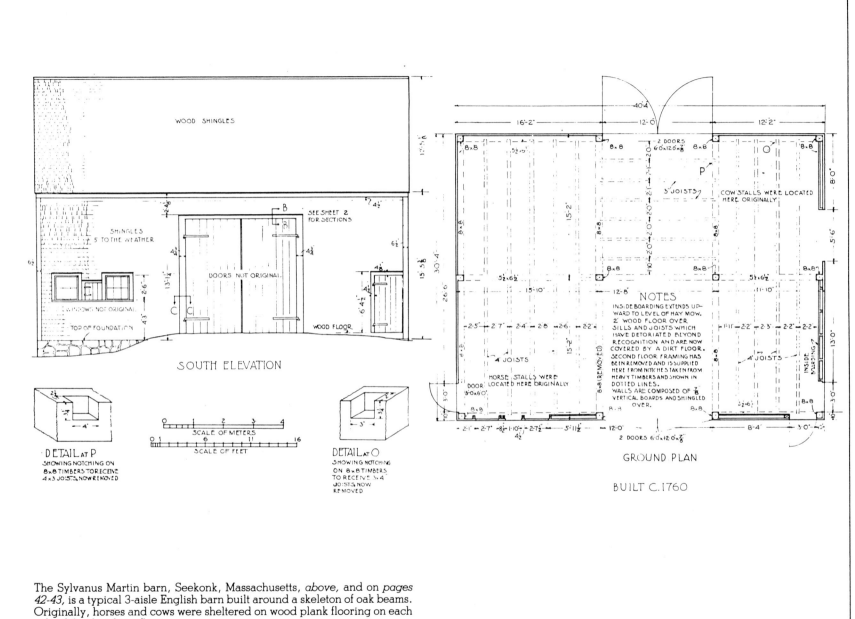

WOOD SHINGLES

SHINGLES
5 TO THE WEATHER

SEE SHEET 2
FOR SECTIONS

DOORS NOT ORIGINAL

WINDOWS NOT ORIGINAL

TOP OF FOUNDATION

WOOD FLOOR

SOUTH ELEVATION

SCALE OF METERS

SCALE OF FEET

DETAIL AT P
SHOWING NOTCHING ON
8×8 TIMBERS TO RECEIVE
4×3 JOISTS NOW REMOVED

DETAIL AT O
SHOWING NOTCHING
ON 8×8 TIMBERS
TO RECEIVE 3×4
JOISTS NOW
REMOVED

2 DOORS
6·0×12·0×⅞

3″ JOISTS

COW STALLS WERE LOCATED
HERE ORIGINALLY

NOTES
INSIDE BOARDING EXTENDS UP-
WARD TO LEVEL OF HAY MOW.
2″ WOOD FLOOR OVER
SILLS AND JOISTS WHICH
HAVE DETORIATED BEYOND
RECOGNITION AND ARE NOW
COVERED BY A DIRT FLOOR.
SECOND FLOOR FRAMING HAS
BEEN REMOVED AND IS SUPPLIED
HERE FROM NOTCHES TAKEN FROM
HEAVY TIMBERS AND SHOWN IN
DOTTED LINES.
WALLS ARE COMPOSED OF ⅞
VERTICAL BOARDS AND SHINGLED
OVER.

4 JOISTS

HORSE STALLS WERE
LOCATED HERE ORIGINALLY
DOOR
3·0×6·0′

2 DOORS 6·0×12·0×⅞

GROUND PLAN

BUILT C. 1760

The Sylvanus Martin barn, Seekonk, Massachusetts, *above,* and on *pages 42-43,* is a typical 3-aisle English barn built around a skeleton of oak beams. Originally, horses and cows were sheltered on wood plank flooring on each side of the threshing floor.

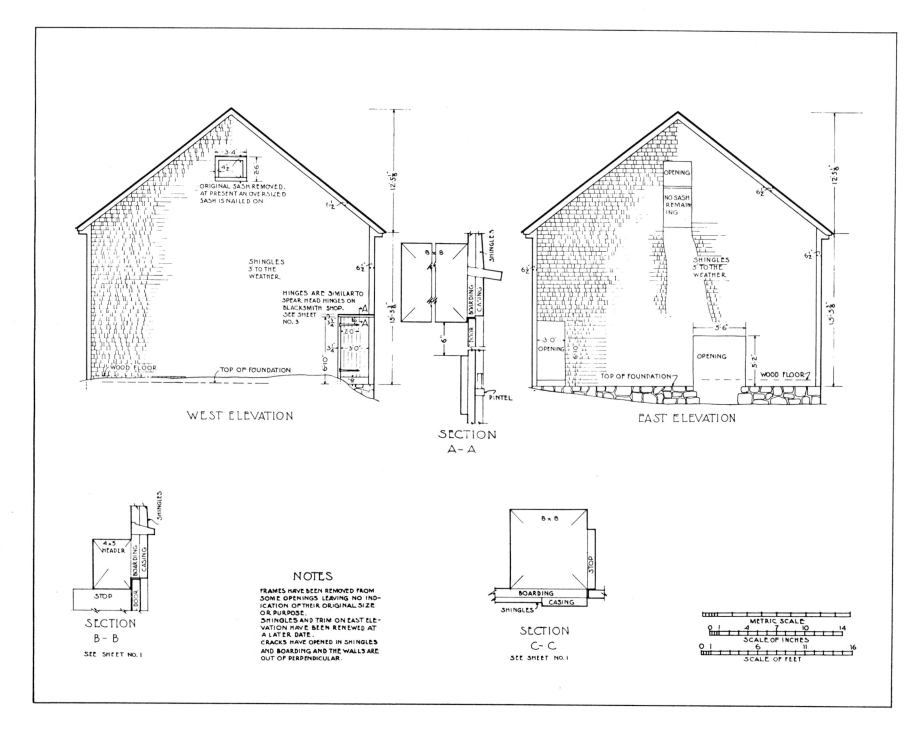

WEST ELEVATION

-3'-4"-
4'-2"
2'-6"
ORIGINAL SASH REMOVED. AT PRESENT AN OVERSIZED SASH IS NAILED ON

SHINGLES 5 TO THE WEATHER

HINGES ARE SIMILAR TO SPEAR HEAD HINGES ON BLACKSMITH SHOP. SEE SHEET NO. 3

WOOD FLOOR

TOP OF FOUNDATION

6'-2"

3¼"
2'-0"
3¼"
3'-0"
6'-10"

12'-5⅝"
15'-5⅝"

SECTION A-A

8 × 8

BOARDING

CASING

SHINGLES

DOOR

6"

PINTEL

EAST ELEVATION

OPENING

NO SASH REMAINING

SHINGLES 5 TO THE WEATHER

6'-2"

6'-2"

3'-0" OPENING

6'-10"

5'-6"

OPENING

5'-2"

TOP OF FOUNDATION

WOOD FLOOR

12'-5⅝"
15'-5⅝"

SECTION B-B

SHINGLES

4×5 HEADER

BOARDING

CASING

DOOR

STOP

SEE SHEET NO. 1

NOTES

FRAMES HAVE BEEN REMOVED FROM SOME OPENINGS LEAVING NO INDICATION OF THEIR ORIGINAL SIZE OR PURPOSE.
SHINGLES AND TRIM ON EAST ELEVATION HAVE BEEN RENEWED AT A LATER DATE.
CRACKS HAVE OPENED IN SHINGLES AND BOARDING AND THE WALLS ARE OUT OF PERPENDICULAR.

SECTION C-C

8 × 8

BOARDING

SHINGLES

CASING

STOP

SEE SHEET NO. 1

METRIC SCALE
0 1   4   7   10   14
SCALE OF INCHES
0 1   6   11   16
SCALE OF FEET

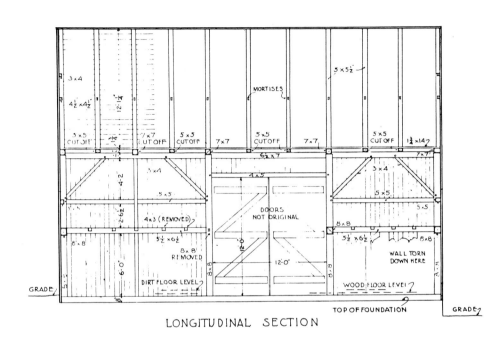

3×4

4½×4½

12'-1¾

5×5½

MORTISES

5×5
CUT OFF

7×7

5×5
CUTOFF

7×7

5×5
CUTOFF

7×7

5×5
CUT OFF

1½×14½

6½×7

7×7

3×4

4×2

5×5

DOORS
NOT ORIGINAL

3×4

5×5

2×6½

4×3 (REMOVED)

5½×6½

8×8
REMOVED

8×8

12'-0

12'-0

5.5

8×8

5½×6½

8×8

WALL TORN
DOWN HERE

DIRT FLOOR LEVEL

6'-0

8×8

WOOD FLOOR LEVEL

GRADE

TOP OF FOUNDATION

GRADE

LONGITUDINAL SECTION

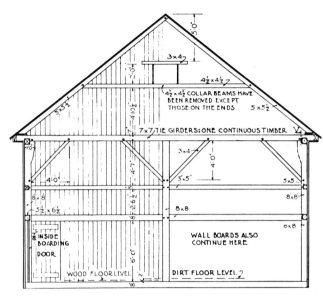

5'-0

3×4½

3'-0

4½×4½

5½×5½

4½×4½ COLLAR BEAMS HAVE
BEEN REMOVED EXCEPT
THOSE ON THE ENDS

5×5½

3'-0

4'-10½

7×7 TIE GIRDERS: ONE CONTINUOUS TIMBER

4'-7

3×4

4'-0

5×5

5×5

8'-8

8×8

8'-2-6½

5½×6½

8×8

8×8

INSIDE
BOARDING

DOOR

WALL BOARDS ALSO
CONTINUE HERE

6'-0

WOOD FLOOR LEVEL

DIRT FLOOR LEVEL

8'-0

GRADE

TRANSVERSE SECTION

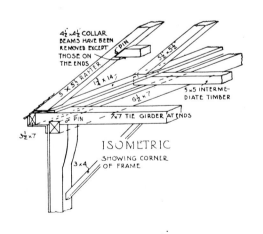

4½×4½ COLLAR
BEAMS HAVE BEEN
REMOVED EXCEPT
THOSE ON
THE ENDS

PIN

5½×5½

5×5½ RAFTER

1½×14½

6½×7

5×5 INTERME-
DIATE TIMBER

PIN

7×7 TIE GIRDER AT ENDS

3½×7

ISOMETRIC
SHOWING CORNER
OF FRAME

3×4

NOTES

REMOVED TIE GIRDERS WERE CUT
AT V AS SHOWN ON TRANSVERSE SECTION.
THE FRAME AND BOARDING ARE OF
OAK; TRIM IS OF WHITE PINE.
BEAMS HAVE BEEN REMOVED
WHERE SO INDICATED AND, IN SOME
INSTANCES, WIRE CABLES WERE
SUBSTITUTED.

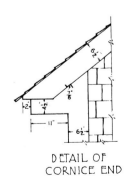

6½

1½
7/8

4½

½

11"

6½

DETAIL OF
CORNICE END

METRIC SCALE

0   1   2   3   4

SCALE OF FEET FOR DETAIL

0 1   6   11   16

SCALE OF FEET

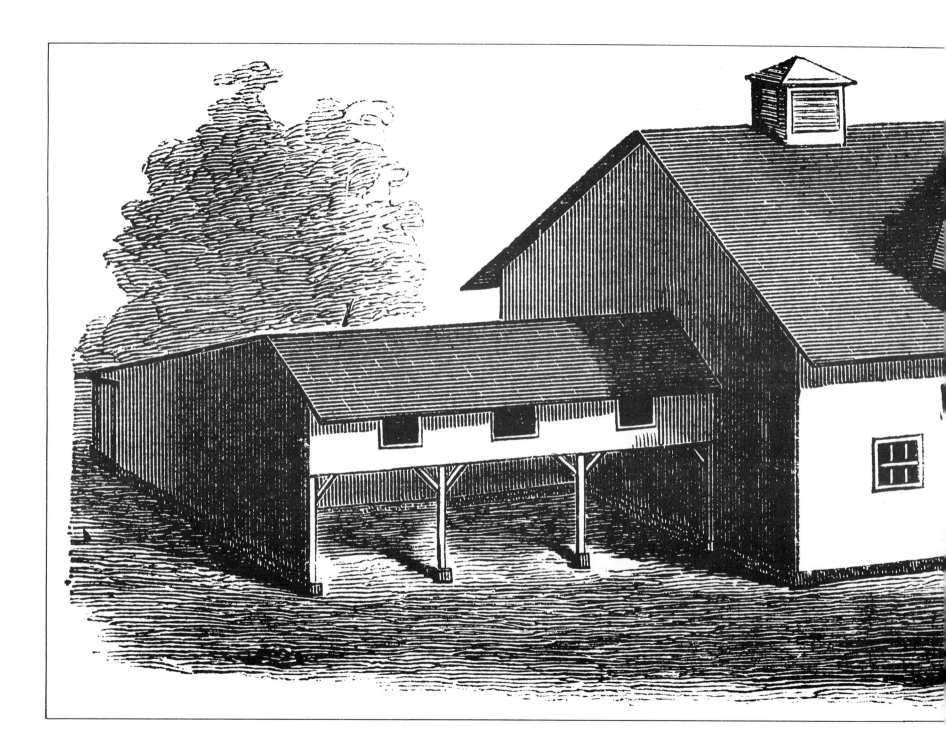

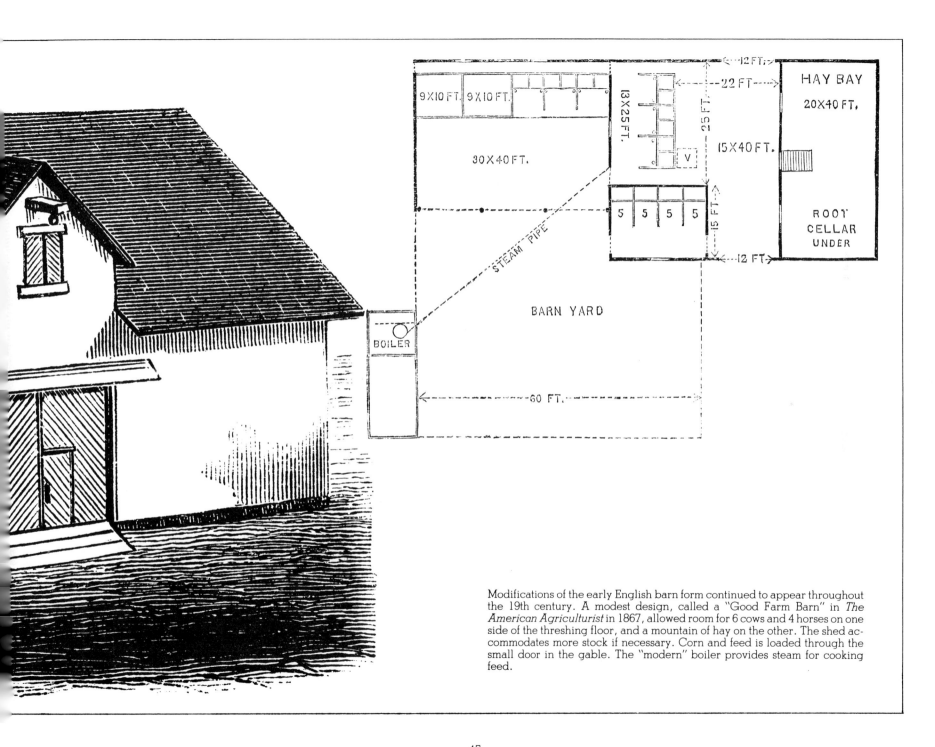

Modifications of the early English barn form continued to appear throughout the 19th century. A modest design, called a "Good Farm Barn" in *The American Agriculturist* in 1867, allowed room for 6 cows and 4 horses on one side of the threshing floor, and a mountain of hay on the other. The shed accommodates more stock if necessary. Corn and feed is loaded through the small door in the gable. The "modern" boiler provides steam for cooking feed.

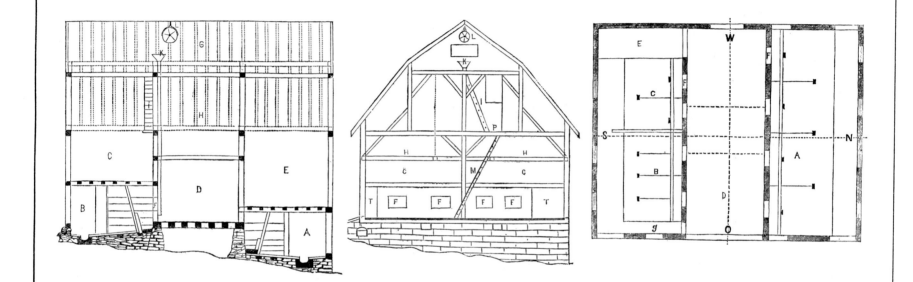

German and English influences combine in this barn built in Potter County, Pennsylvania, in the 1860s. The driveway between mows is typically English while the gambrel roof and the use of the cellar for livestock is more often found in German-American barns.

A three-story barn. Rather than adding outbuildings, many 19th-century builders preferred to add floors. From the front, this "triple-decker" looks like an English barn.

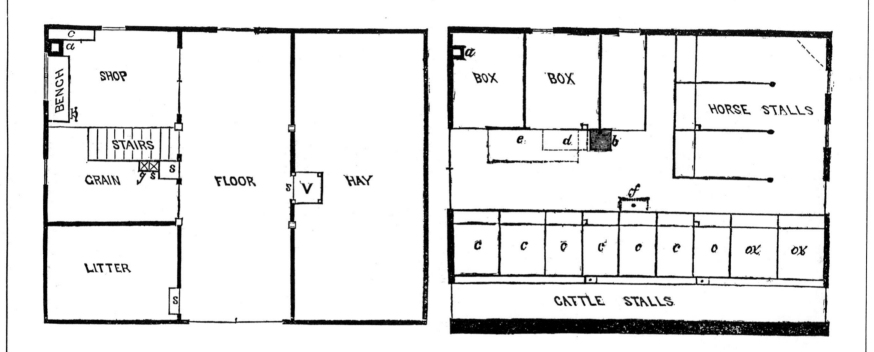

A three-story barn. A shop and granary occupy one-third of the main floor.
Chutes for feed and bedding descend from granary and loft to stock floors.
The door to the cattle stalls is through the gable end at ground level.

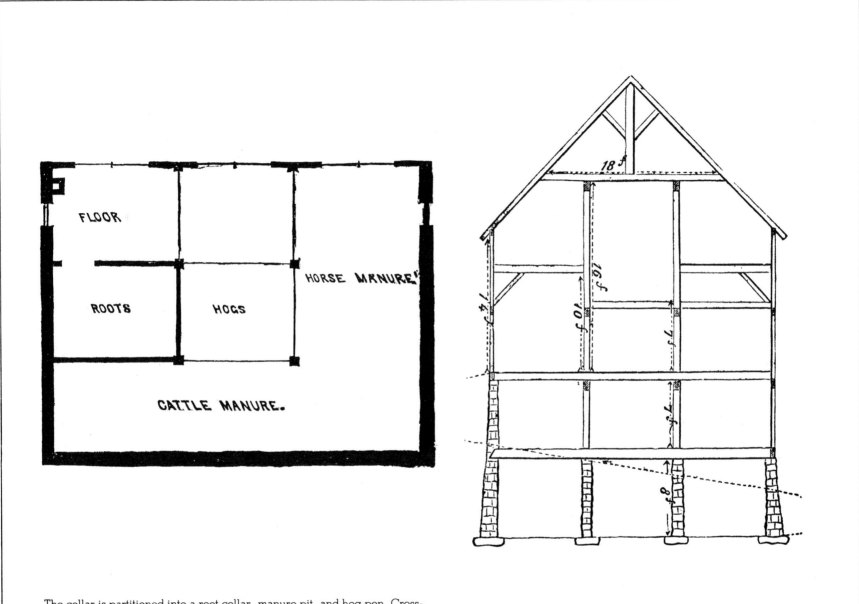

The cellar is partitioned into a root cellar, manure pit, and hog pen. Cross-section shows framing and slope of ground. Topmost tie beams can be planked and used for drying corn in the heat of the loft.

49

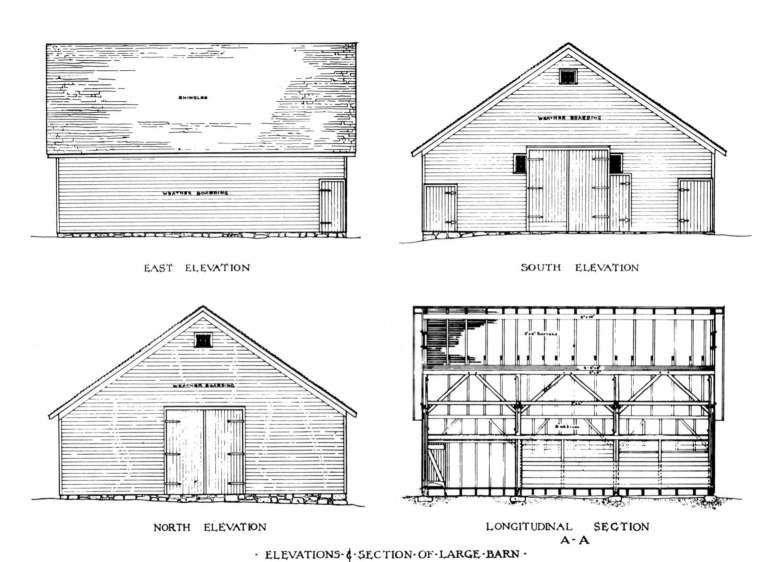

EAST ELEVATION

SOUTH ELEVATION

NORTH ELEVATION

LONGITUDINAL SECTION
A-A

· ELEVATIONS · & · SECTION · OF · LARGE · BARN ·

The Jacob Zabriskie barn shown on this and the facing page was built late in the period of Dutch influence but retains the characteristic Dutch gable-end doors and wide roof.

50

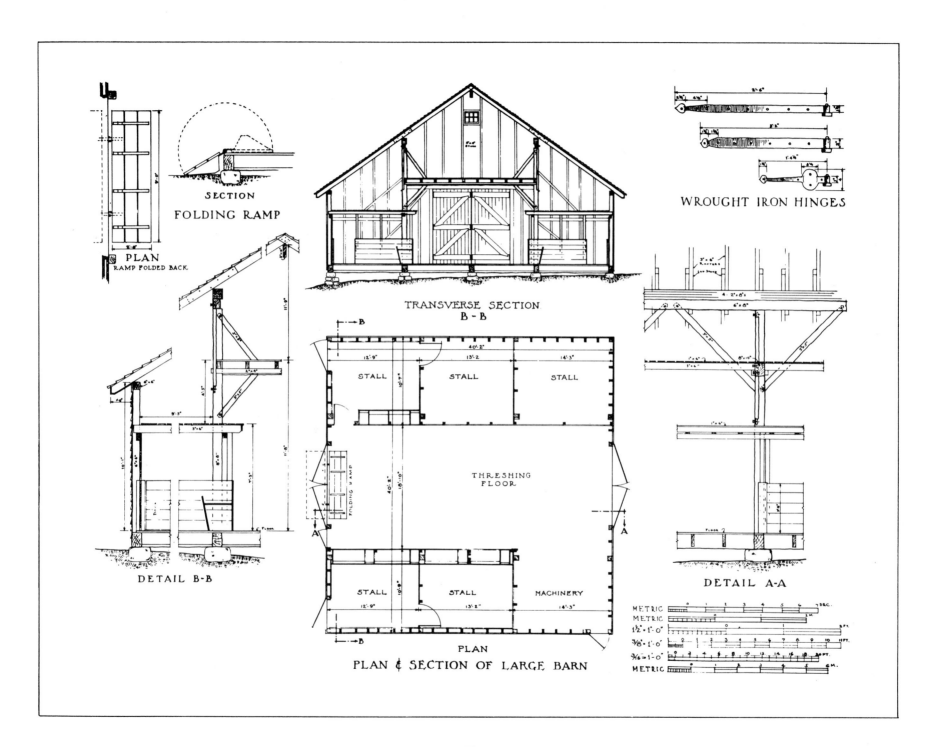

SECTION
FOLDING RAMP

PLAN
RAMP FOLDED BACK.

DETAIL B-B

TRANSVERSE SECTION
B - B

WROUGHT IRON HINGES

DETAIL A-A

STALL

STALL

STALL

THRESHING
FLOOR

FOLDING RAMP

STALL

STALL

MACHINERY

PLAN

PLAN & SECTION OF LARGE BARN

METRIC
METRIC
1½" = 1'- 0"
3/8" = 1'- 0"
3/16" = 1'- 0"
METRIC

51

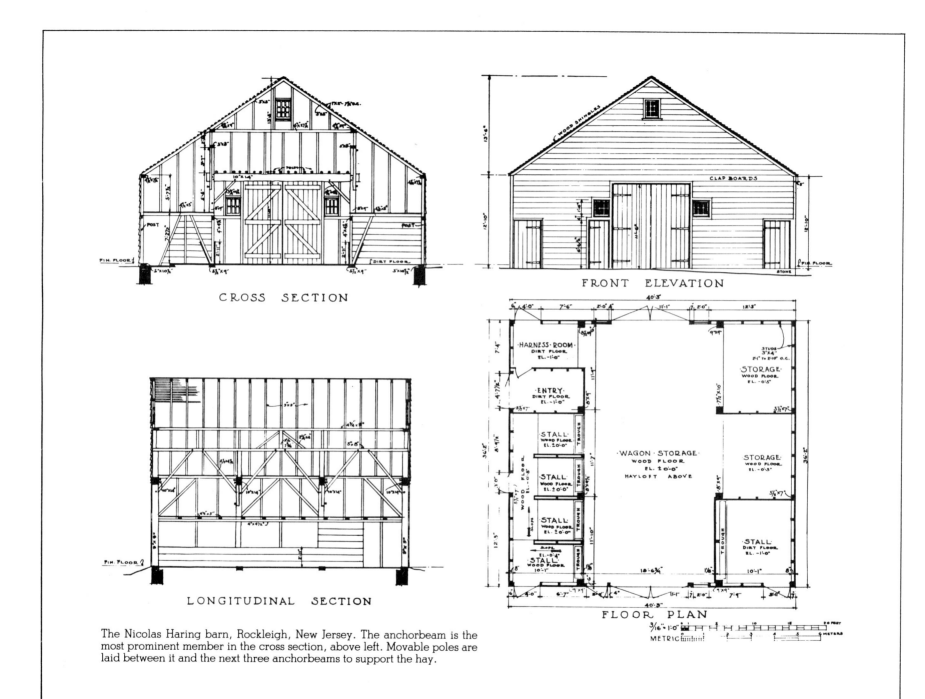

CROSS SECTION

FRONT ELEVATION

LONGITUDINAL SECTION

FLOOR PLAN

The Nicolas Haring barn, Rockleigh, New Jersey. The anchorbeam is the most prominent member in the cross section, above left. Movable poles are laid between it and the next three anchorbeams to support the hay.

52

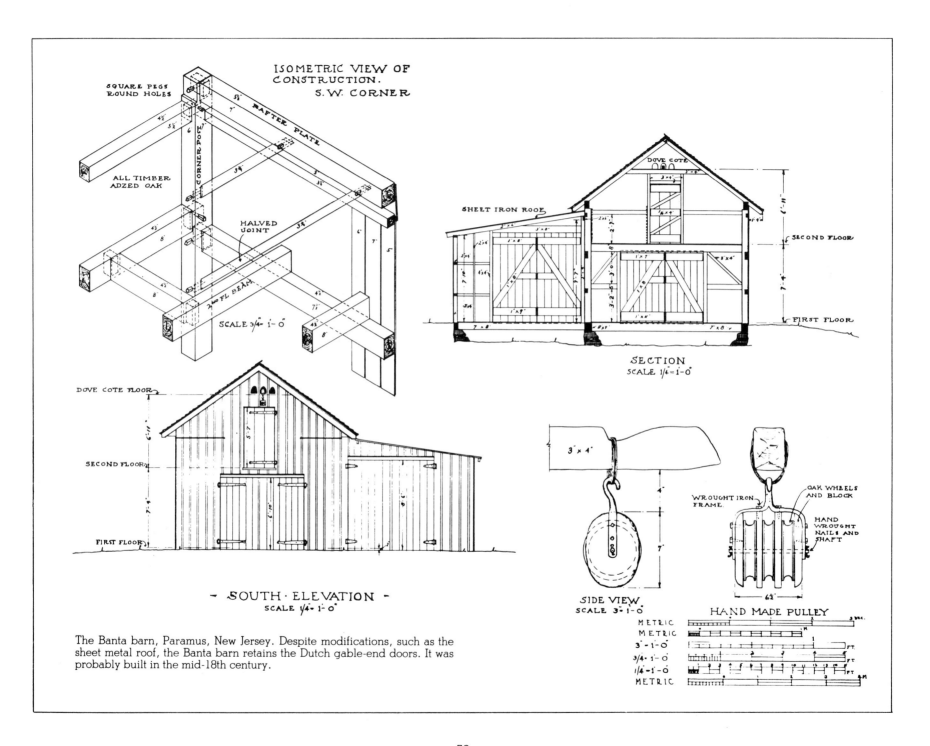

ISOMETRIC·VIEW·OF
CONSTRUCTION.
S.W. CORNER

SQUARE PEGS
ROUND HOLES

ALL TIMBER
ADZED OAK

CORNER POST

RAFTER PLATE

HALVED
JOINT

2ND FL BEAM

SCALE 3/4" = 1'-0"

SHEET IRON ROOF

DOVE COTE

SECOND FLOOR

FIRST FLOOR

SECTION
SCALE 1/4" = 1'-0"

DOVE COTE FLOOR

SECOND FLOOR

FIRST FLOOR

- SOUTH · ELEVATION -
SCALE 1/4" = 1'-0"

The Banta barn, Paramus, New Jersey. Despite modifications, such as the
sheet metal roof, the Banta barn retains the Dutch gable-end doors. It was
probably built in the mid-18th century.

3" x 4"

WROUGHT IRON
FRAME.

OAK WHEELS
AND BLOCK

HAND
WROUGHT
NAILS AND
SHAFT

SIDE VIEW
SCALE 3" = 1'-0"

HAND MADE PULLEY

METRIC
METRIC
3" = 1'-0"
3/4" = 1'-0"
1/4" = 1'-0"
METRIC

53

An advertisement for Fairbanks scales in *The American Agriculturist,* 1870. The engraving clearly shows the ramp to the threshing floor, the storage area under the ramp which may conceal a root cellar and a cistern, and the opening to the ground-floor stables.

The New York-based *American Agriculturist* never failed to congratulate its Pennsylvania neighbors for being such competent farmers and barn builders:

An ample barn for the storage of crops and the shelter of stock should be regarded as a necessary investment of capital in all farming in the Northern and Eastern States. This is better understood in Pennsylvania than in any other part of the country, and the barn that bears the name of the State is, in many respects, a model.

The most distinguishing features of the Pennsylvania barn—or what is also known as a bank or sidehill barn—are the ramp on the uphill side and the forebay on the other. By slicing out a relatively small amount of earth and building into a hillside or banking up a grade on level ground, a farmer gained ground-level access to the threshing floor and mows from one side and to the stables below from the other. The forebay is a cantilevered extension of the threshing floor projecting at least 6 feet beyond the stable doors below. It forms a protected area in front of the doors and runs the length of the barn. Some forebays are supported either by extended gable-end walls or by pillars.

Frame and masonry construction rapidly replaced the log and stone barns early in the 19th century. The gable ends were likely to be built of brick or stone while the frame in between was made of heavy oak beams sheathed with vertical rough boards and supported by a stone foundation. The brick ventilator patterns on many barns in Pennsylvania and Ohio add to their whimsical charm.

Basement barns were not without their critics in the final third of the 19th century when sanitation was a growing concern. The natural consequence of having masonry stable walls below grade was the ever-present condensation and mildew which, when combined with the vapors of decomposing urine in the ground and the formation of saltpeter on the stone walls, made for a generally unhealthy atmosphere for both the livestock and for the crops stored on the floor above.

Notwithstanding such valid criticism, the basic Pennsylvania barn form survived. Gravel and pitch and later concrete floors were laid where stone or dirt alone had sufficed; ventilation was improved.

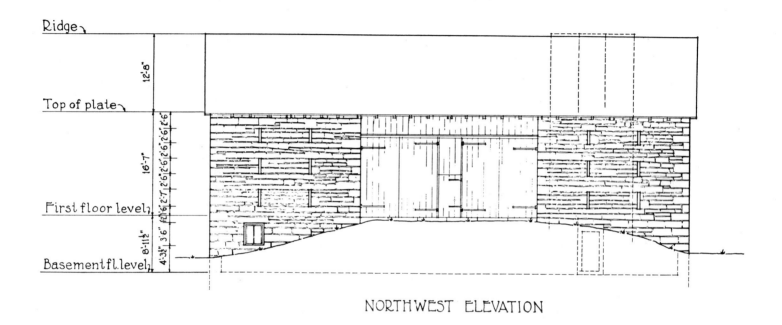

Ridge

12'-8"

Top of plate

16'-7"

2'-6" 2'-6" 2'-6" 2'-6" 2'-7" 2'-6"

First floor level

8'-11½"

3'-6"

4'-3½"

Basement fl. level

NORTHWEST ELEVATION

David Stauffer barn, Butler County, Pennsylvania. Stone barns like this are a common sight in southeastern Pennsylvania. Double doors provide turn-around space for a wagon team. Dotted lines at right indicate a silo recently erected. Vertical slits ventilate the hay mows. Window at lower left opens on basement.

Scale ⅛"=1'-0"   0   5   10 feet

Metric   1   0   1   2   3   4   5 metres

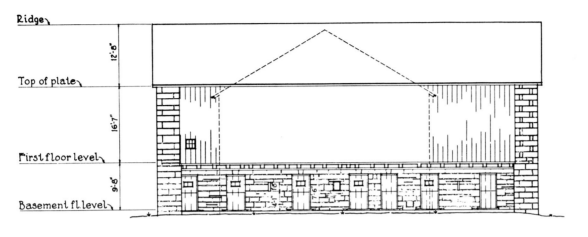

SOUTHEAST ELEVATION

Ridge

Top of plate

First floor level

Basement fl. level

12'-8"

16'-7"

9'-8"

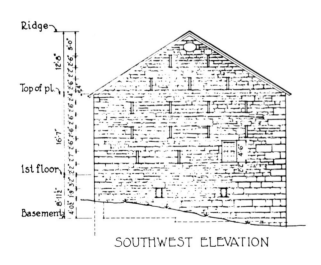

SOUTHWEST ELEVATION

Ridge

Top of pl.

1st floor

Basement

12'-8"

5'-0"

16'-7"

8'-11½"

General Notes
Stone masonry is of local sandstone: The dotted lines laid over northeast and southeast elevations represent additions made to the original building: Inscription tablets in gable ends have been removed: Doors are batten dutch doors of one inch boards.

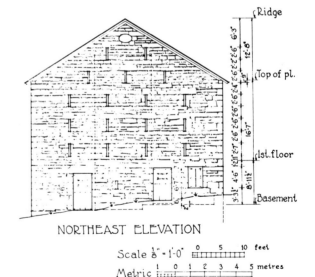

NORTHEAST ELEVATION

Ridge

Top of pl.

1st floor

Basement

12'-8"

6'-3"

16'-7"

8'-11½"

Scale ⅛" = 1'-0"   0   5   10   feet

Metric   1 0 1 2 3 4 5 metres

The downhill side of the Stauffer barn (southeast elevation) is the front of the building. The stable doors are set back about 6' under the forebay. Support for the forebay comes from gable-end walls and cantilevered 10" x 10" first-floor beams.

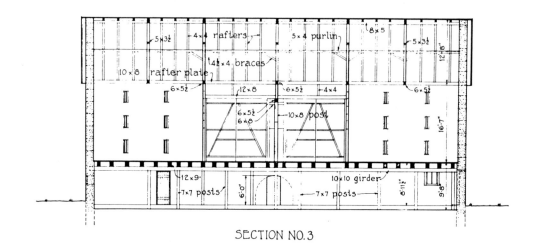

5 x 3½   4 x 4 rafters   5 x 4 purlin   8 x 5   5 x 3½

12'-8"

10 x 8 rafter plate   4½ x 4 braces

6 x 5½   12 x 8   6 x 5½   4 x 4   6 x 5½

6 x 5½   10 x 8 post
6 x 8

16'-1"

12 x 9   10 x 10 girder
7 x 7 posts   6'-0"   7 x 7 posts   8'-11½"   9'-8"

SECTION NO. 3

▨ Stone
▬ Wood

General Notes

The barn is framed of hand hewn oak fastened with wooden pins: Plank are laid on movable timbers to form an oats mow: The first floor framing is cantilevered over basement wall to carry rear wall framing of first floor.

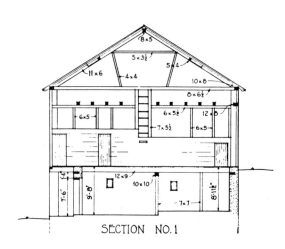

8 x 5
5 x 3½
11 x 6   4 x 4   5 x 4   10 x 8
8 x 6½
6 x 5   6 x 5½   12 x 8
7 x 5½   6 x 5
12 x 9
7'-6"   9'-8"   10 x 10   8'-11½"
7 x 7

SECTION NO. 1

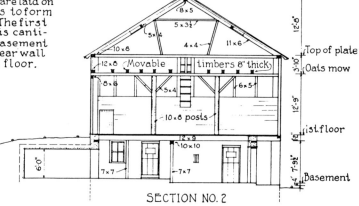

Ridge
8 x 5
5 x 3½   12'-8"
5 x 4   4 x 4   11 x 6
10 x 8   Top of plate
12 x 8 Movable timbers 8" thick   3'-10" Oats mow
8 x 8   6 x 5½
5 x 4   12'-9"
10 x 8 posts
6'-0"   12 x 9   1st floor
10 x 10   6'-0"   7'-9½"
7 x 7   7 x 7   Basement   4'-4"

SECTION NO. 2

Scale ⅛" = 1'-0"   0   5   10 feet
Metric   1 0 1 2 3 4 5 metres

Section No. 1 cuts through the Stauffer barn at the far threshing-floor wall. Section No. 2 centers on threshing floor. Section No. 3 runs through the midline.

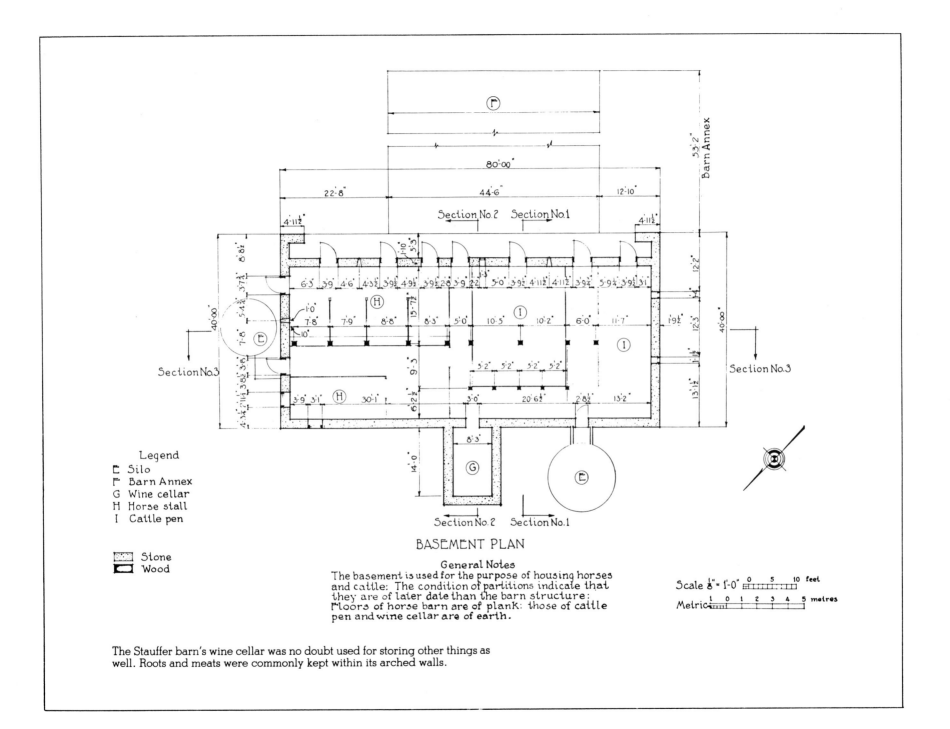

Section No.2   Section No.1

Barn Annex

Section No.3

Section No.3

Section No.2   Section No.1

Legend
E Silo
F Barn Annex
G Wine cellar
H Horse stall
I Cattle pen

▦ Stone
▭ Wood

BASEMENT PLAN

General Notes
The basement is used for the purpose of housing horses
and cattle. The condition of partitions indicate that
they are of later date than the barn structure.
Floors of horse barn are of plank: those of cattle
pen and wine cellar are of earth.

Scale ⅛" = 1'-0"   0   5   10 feet
Metric   1 0 1 2 3 4 5 metres

The Stauffer barn's wine cellar was no doubt used for storing other things as
well. Roots and meats were commonly kept within its arched walls.

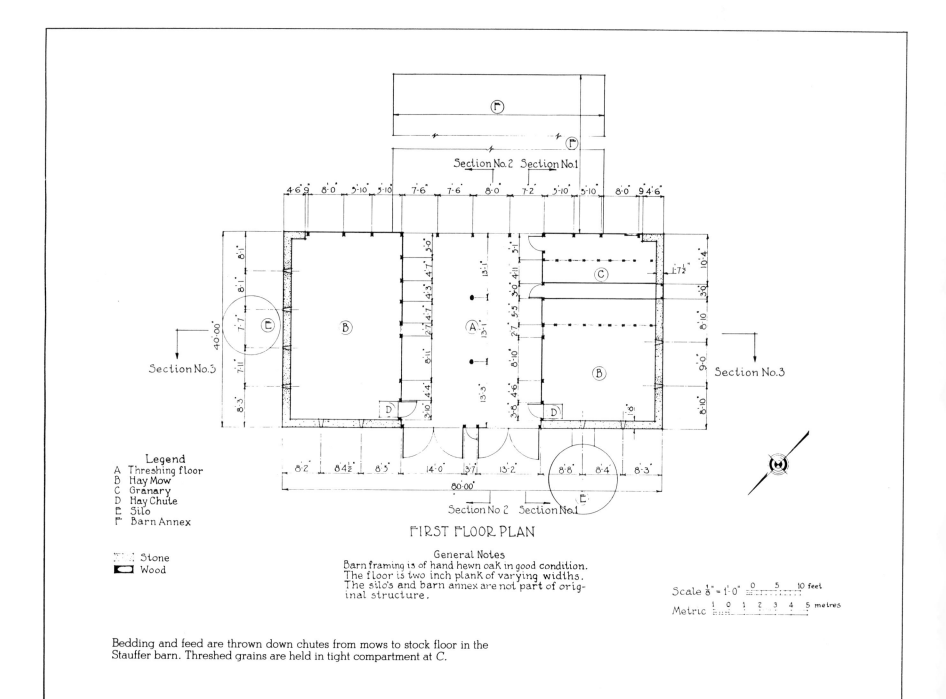

Section No. 2   Section No. 1

F

4'-6" 9'   8'-0"   5'-10"   5'-10"   7'-6"   7'-6"   8'-0"   7'-2"   5'-10"   5'-10"   8'-0"   9'-4" 6'

Section No. 3

Section No. 3

40'-00"

B

A

B

C

D

D

E

E

8'-2"   8'-4½"   8'-5"   14'-0"   13'-7"   13'-2"   8'-8"   8'-4"   8'-3"

80'-00"

Section No. 2   Section No. 1

Legend
A Threshing floor
B Hay Mow
C Granary
D Hay Chute
E Silo
F Barn Annex

Stone
Wood

FIRST FLOOR PLAN

General Notes
Barn framing is of hand hewn oak in good condition.
The floor is two inch plank of varying widths.
The silo's and barn annex are not part of orig-
inal structure.

Scale ⅛" = 1'-0"   0   5   10 feet

Metric   1 0 1 2 3 4 5 metres

Bedding and feed are thrown down chutes from mows to stock floor in the
Stauffer barn. Threshed grains are held in tight compartment at C.

Pennsylvania barn, 1865. A gable roof runs perpendicular to the main ridge and provides additional space in the loft.

The threshold of the wagon doors is 17′ above the stable floor. The extension of the wagonway into the hill reduces the vertical drop by 9′, sufficient to allow the easy ascent by a team pulling a loaded wagon.

Overleaf: Space is used efficiently below the wagonway. A 10′ x 40′ wagon house has doors at each end and passages to the stable on one side and to a root cellar on the other. A 20,000 gallon cistern abuts the root cellar. The base of the reservoir is 18″ above the level of the stable floor.

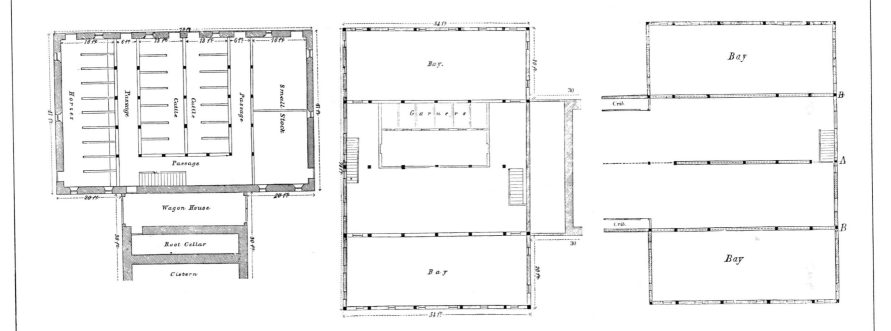

The stalls in the basement, *above left,* are partitioned by oak slats, 2″ wide and 2″ apart, supported by 4″ x 4″ studs. The floors are paved with square wood blocks—end pieces saved during the barn's erection—cut to 6″ lengths. They are laid together on a 4″-deep bed of lime to prevent decay. Gaps between the blocks were chinked before the entire floor was covered with pitch, sand, and finally, straw. Gangways are either planked or gravel-filled and saturated with pitch. "Garners" in the middle floor, *center,* store grain received through a grate in the threshing floor above. The floor stops at the bays which extend from the ceiling of the stables to the rafters. Wagons are unloaded on the threshing floor, *right,* by a horse-powered hay fork. The load is thrown off into the bays on either side.

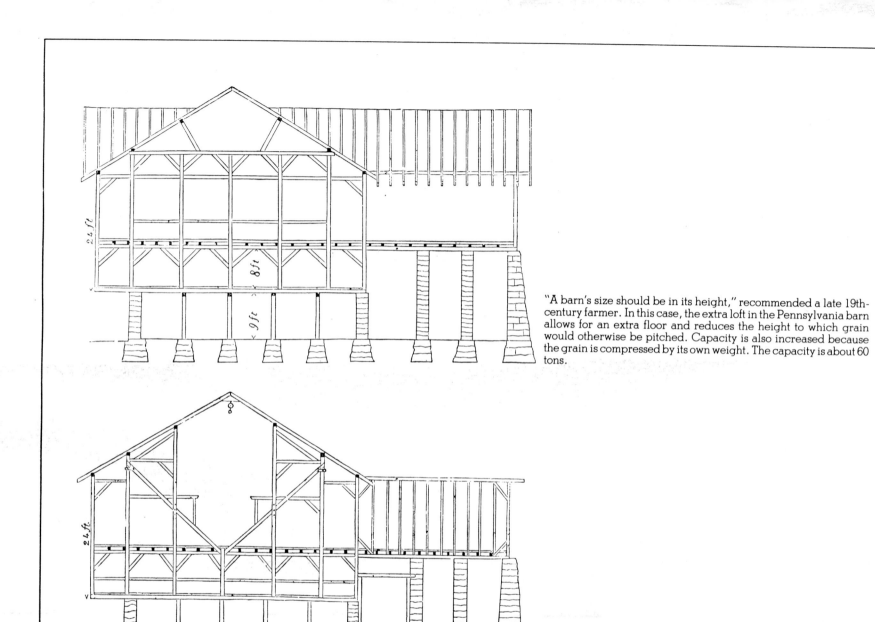

"A barn's size should be in its height," recommended a late 19th-century farmer. In this case, the extra loft in the Pennsylvania barn allows for an extra floor and reduces the height to which grain would otherwise be pitched. Capacity is also increased because the grain is compressed by its own weight. The capacity is about 60 tons.

The cross section at the center of the threshing floor shows that the cross-timbers were to be constructed to permit the hayfork to swing freely across the mows.

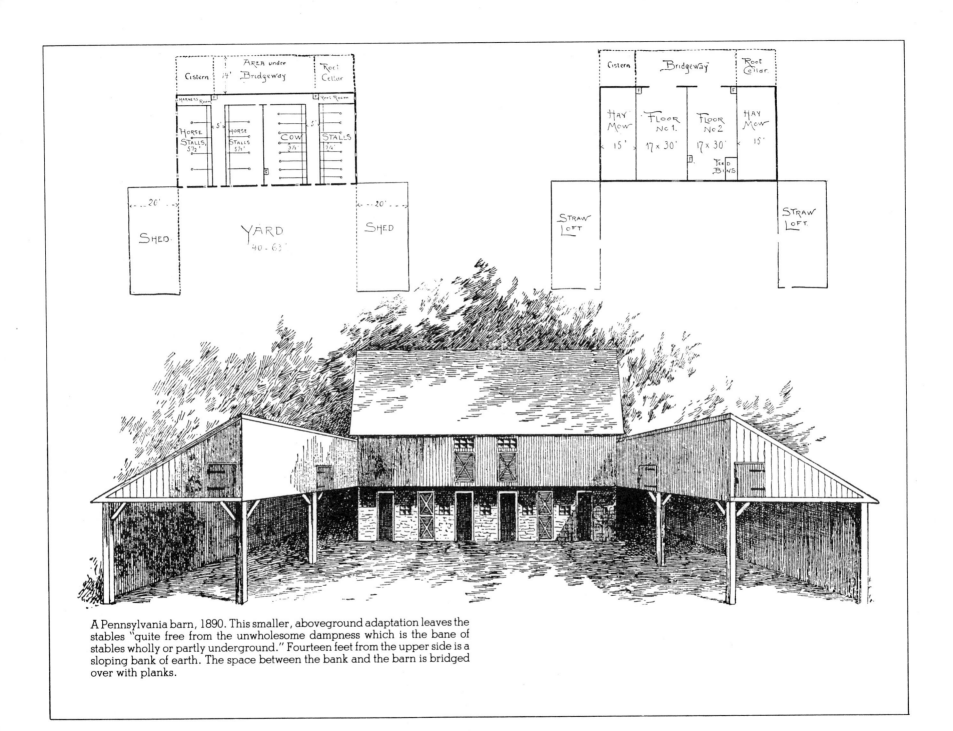

Cistern · AREA under 14′ Bridgeway · Root Cellar

HARNESS Room · Root Room

HORSE STALLS 5½′ · 5′ · HORSE STALLS 5½′ · COW 3½ · STALLS 3½ · 5′

20′ · SHED · YARD 40 - 63′ · 20′ · SHED

Cistern · Bridgeway · Root Cellar

HAY MOW 15′ · FLOOR No 1. 17 x 30′ · FLOOR No 2 17 x 30′ · HAY MOW 15′ · FEED BINS

STRAW LOFT · STRAW LOFT.

A Pennsylvania barn, 1890. This smaller, aboveground adaptation leaves the stables "quite free from the unwholesome dampness which is the bane of stables wholly or partly underground." Fourteen feet from the upper side is a sloping bank of earth. The space between the bank and the barn is bridged over with planks.

65

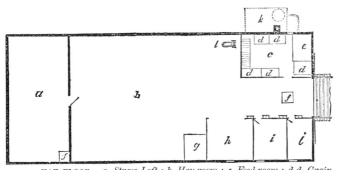

HAY FLOOR.—*a, Straw Loft ; b, Hay-room ; c, Feed-room ; d d, Grain-bins ; e, Steaming Vat ; f f, Hatches ; g, Water-tank ; h, Tool-room ; i, Work-shop ; j, Bed-room ; k, Boiler-room ; l, Hay-cutter.*

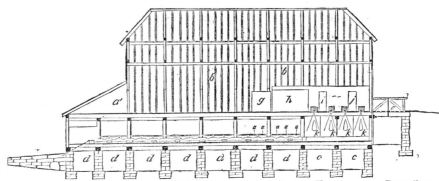

LONGITUDINAL SECTION.—*a a, Cattle Stalls ; a', Straw Loft ; b b, Hay-room ; c c, Root-cellar ; d d, Manure Cellar ; e, Bridge ; g, Water-tank ; h, Tool-room ; i, Door to Work-shop ; j, Do. to Bed-room ; k k, Horse Stalls.*

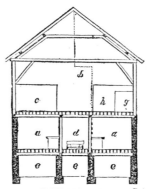

CROSS-SECTION.—*a a, Cat-tle Stalls ; b, Hay-room ; c, Feed-room ; d, Passage with Car for Feed ; e e, Manure Cellar ; g, Water-tank ; h, Tool-room.*

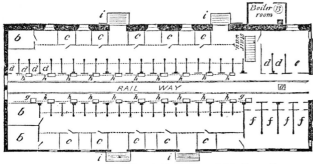

CATTLE FLOOR.—*a, Trap to Root-cellar ; b b b, Loose Boxes ; c c, Calf Pens ; d d, Cow Stalls ; e, Ox Stalls ; f f, Horse Stalls ; g g, Water-tubs ; h h, Watering Troughs ; i i, Slopes to Enter.*

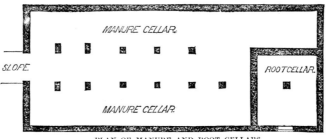

PLAN OF MANURE AND ROOT CELLARS.

The Ogden barn, Orange County, New York, 1880s. The barn on the opposite page and in the plans at the left takes advantage of a ramp to the upper floor, but, unlike those of a Pennsylvania barn, its doors are located at the gable end. It was originally built for $7500.

67

Ohio barn, 1872. From the forebay side it resembles a Pennsylvania barn, but, like the Ogden barn, it is ramped at the gable end. A root cellar occupies the space under the ramp.

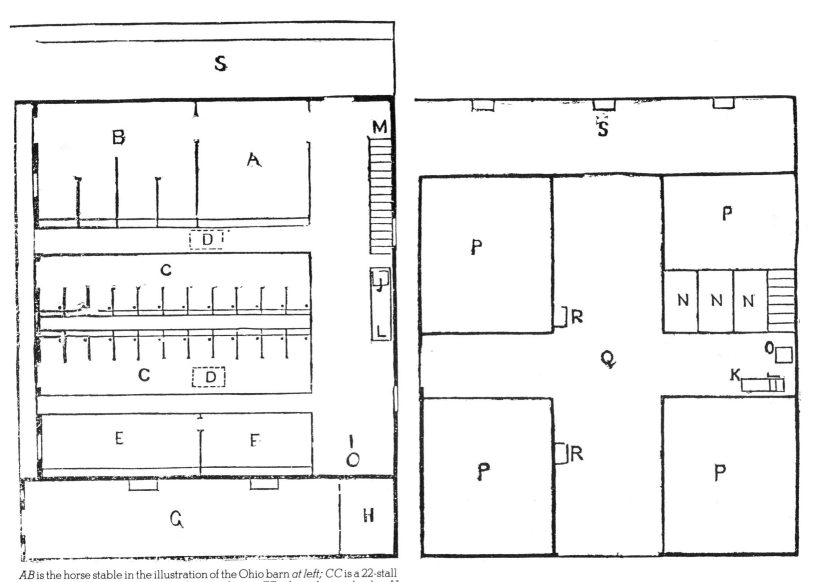

AB is the horse stable in the illustration of the Ohio barn *at left;* CC is a 22-stall stable; DD are ventilators and straw chutes; EF is for calves or lambs; H and I are cistern and pump; J is a feed chute; L is a steam chest. Upper floor, *right,* contains a threshing area and cross-aisle for storing machinery. Granary is at N; bays at P.

Houghton Farm barn, Mountainville, New York, 1878. *Above:* Horse stall is constructed of varnished tongue and groove pine topped with iron bars. *Opposite page, top:* Main floor contains horse power, granary, tool room, sleeping quarters for a hand, a large wagon scale, and hay bays. *Opposite page, bottom:* Ground-floor stables house horses, oxen, and cows. Stable vents ascend to cupola.

A Prince Edward Island, Canada, bank barn, 1886. Measuring 52' x 64', the building was used for fattening sheep and cattle. Its steeply-pitched roof is an excellent snow-shedder. Gutters collect and channel water to holding tanks for watering stock and cooking feed.

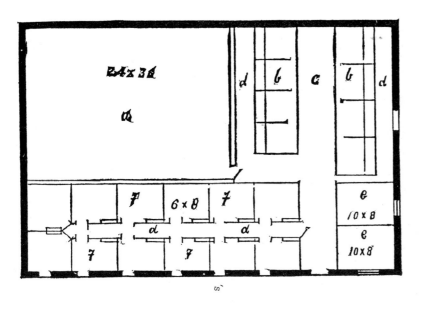

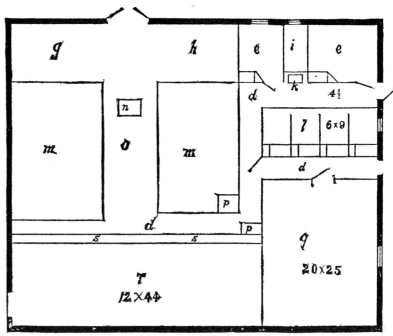

The basement of the Prince Edward Island barn, *left,* contains a 3500-bushel root cellar and pens for cattle and swine. The main floor, *right,* houses horses and sheep, and includes a granary, hay mows, and an equipment room. A hatch in the threshing floor opens to the root cellar.

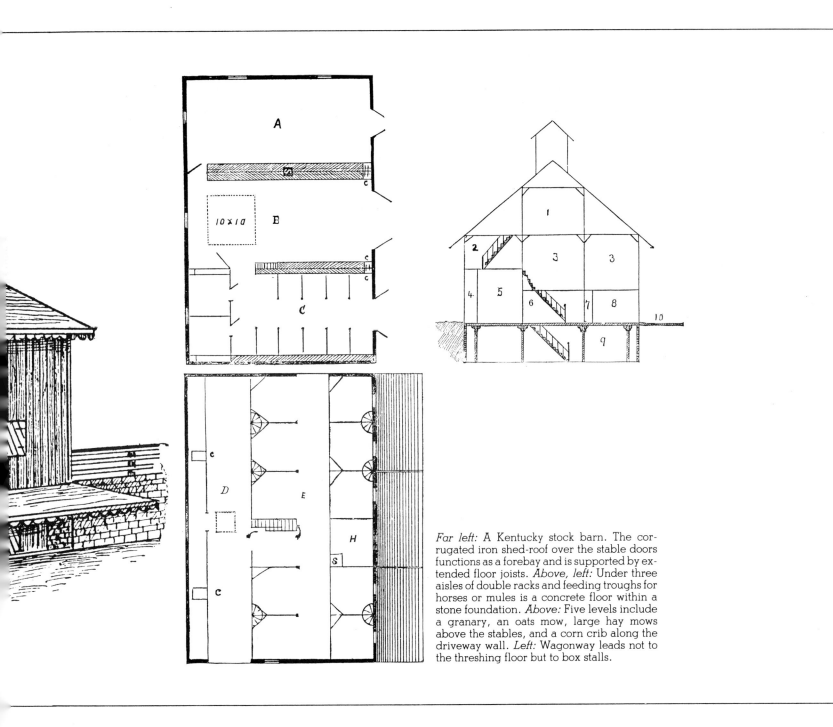

Far left: A Kentucky stock barn. The corrugated iron shed-roof over the stable doors functions as a forebay and is supported by extended floor joists. *Above, left:* Under three aisles of double racks and feeding troughs for horses or mules is a concrete floor within a stone foundation. *Above:* Five levels include a granary, an oats mow, large hay mows above the stables, and a corn crib along the driveway wall. *Left:* Wagonway leads not to the threshing floor but to box stalls.

The octagon barn, be it austere and imposing, or rural gimcrackery, was the most common departure from the ubiquitous, modest rectangle.

experimental barns

The search for heavenly space on earth by 19th century idealists left its legacy in radically new kinds of barns. As early as 1826, the Shakers built a round barn at Hancock, Massachusetts, which for its formal simplicity was unsurpassed in the secular world. At the "hub" is a hollow octagonal vent running from ground level to louvered cupola. Encircling this shaft is a hay storage area 55 feet in diameter and 35 feet deep. This, in turn, is ringed by an upper-level wagon driveway, a middle tier for dairy cattle, and a basement into which manure is dropped through trap doors for collection by carts. The three distinct sections are clearly visible from outside the barn; the cupola is perched atop the windowed clerestory roof of the hay mow, while the great mass of the superstructure is taken up by the 25 feet high stone wall. The Shakers were blessed with a willing and highly productive labor force which enabled them to build such ambitious projects as the round barn and other great bank barns on their several principal communes in

Massachusetts, New Hampshire, Ohio, and New York. One sees in both their more traditional stone and wood barns a Pennsylvania barn writ large. Built into a hillside, Shaker bank barns may have as many as three ground-level entrances and are commodious enough to house all the farm operations that can be performed indoors.

The octagonal and round barns found scattered throughout North America incorporated and refined much of what the Shakers had done earlier. In some cases, for example, the central ventilating shaft was also a chute for feed; once the silo achieved wide acceptance, it seemed logical that the center of the barn be used for such a purpose. One octagonal barn in Pennsylvania was equipped with a tramway that radiated from a central turntable, by means of which a feed car could be run between the feed-room and the various stalls.

Despite their efficiency and pleasant geometric novelty, the round and octagonal barns were still relatively rare on the farms of North America,

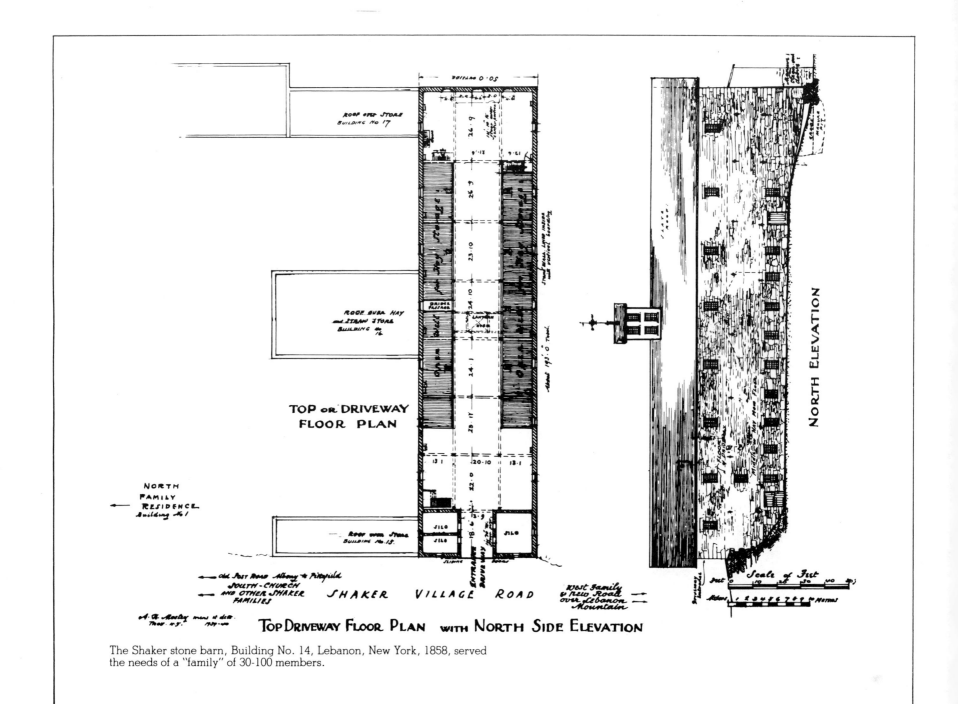

TOP OR DRIVEWAY
FLOOR PLAN

NORTH
FAMILY
RESIDENCE
Building No 1

ROOF OVER STORE
BUILDING NO 17

ROOF OVER HAY
and STRAW STORE
BUILDING No 16

ROOF OVER STORE
BUILDING No 15

SILO
SILO
SILO

SHAKER VILLAGE ROAD

TOP DRIVEWAY FLOOR PLAN WITH NORTH SIDE ELEVATION

NORTH ELEVATION

Scale of Feet

The Shaker stone barn, Building No. 14, Lebanon, New York, 1858, served
the needs of a "family" of 30-100 members.

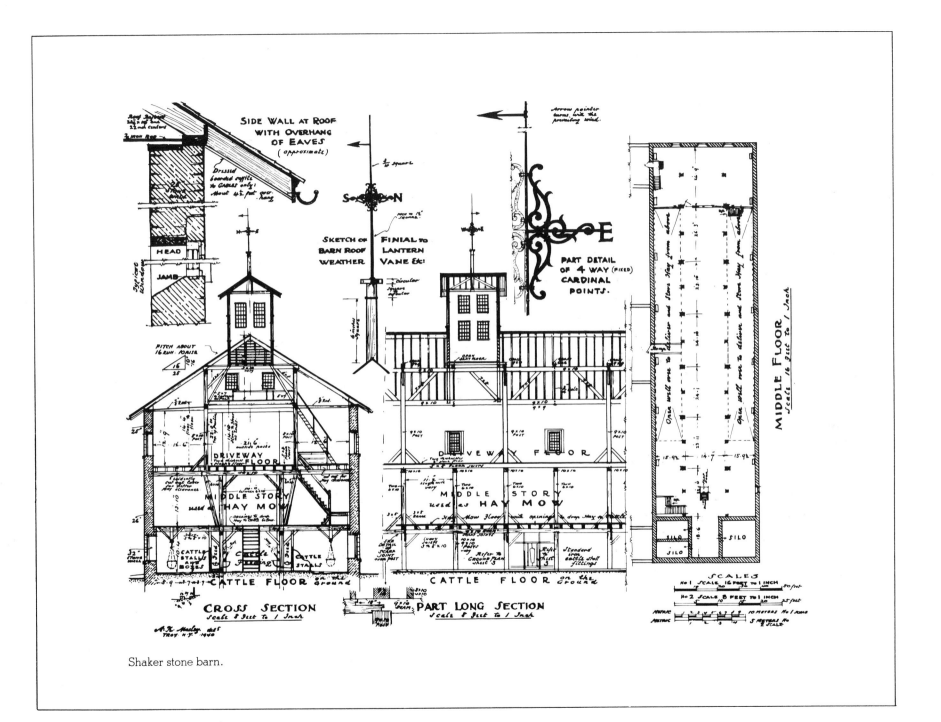

Shaker stone barn.

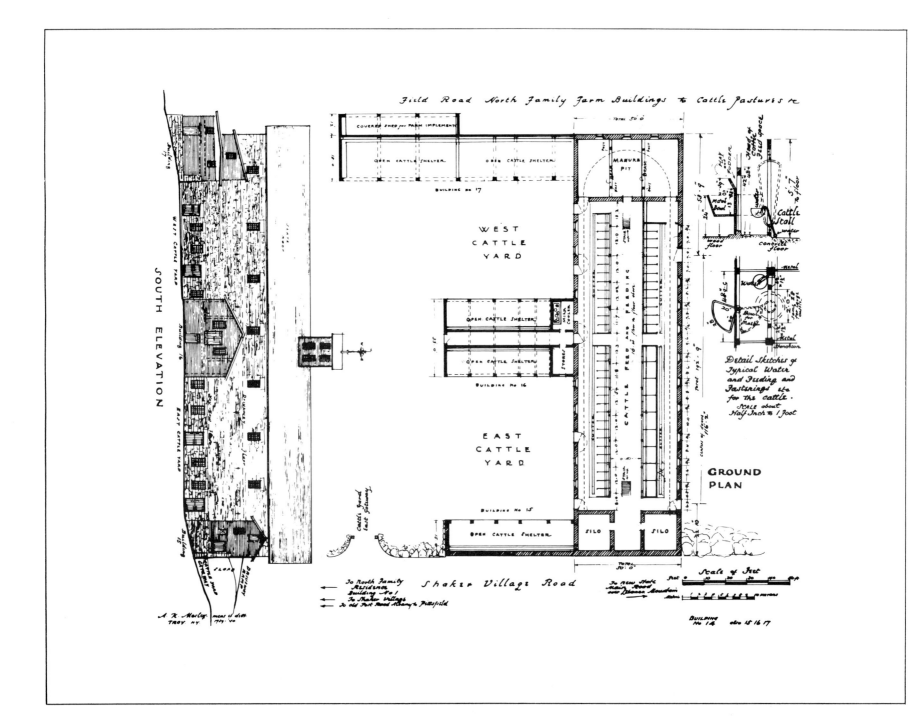

Field Road North Family Farm Buildings to Cattle Pastures &c

COVERED SHED for FARM IMPLEMENT

OPEN CATTLE SHELTER   OPEN CATTLE SHELTER

BUILDING No 17

WEST CATTLE YARD

OPEN CATTLE SHELTER

OPEN CATTLE SHELTER

BUILDING No 16

EAST CATTLE YARD

BUILDING No 15

OPEN CATTLE SHELTER

MANURE PIT

MILK COOLER

CATTLE FEED AND FEEDING

SILO   SILO

Cattle Yard East Gateway

SOUTH ELEVATION

WEST CATTLE YARD

EAST CATTLE YARD

Building 17   Building 16   Building 15

SLOPE

Detail Sketches of Typical Water and Feeding and Fastenings &c for the cattle.
SCALE about Half Inch to 1 Foot

Cattle Stall

GROUND PLAN

Scale of Feet

To North Family Residence Building No 1
To Shaker Village
To old Post Road Albany to Pittsfield

Shaker Village Road

To New State Main Road over Lebanon Mountain

A. K. Morley. meas & delt
Troy N.Y.   1949. '50

BUILDING No 14   also 15 16 17

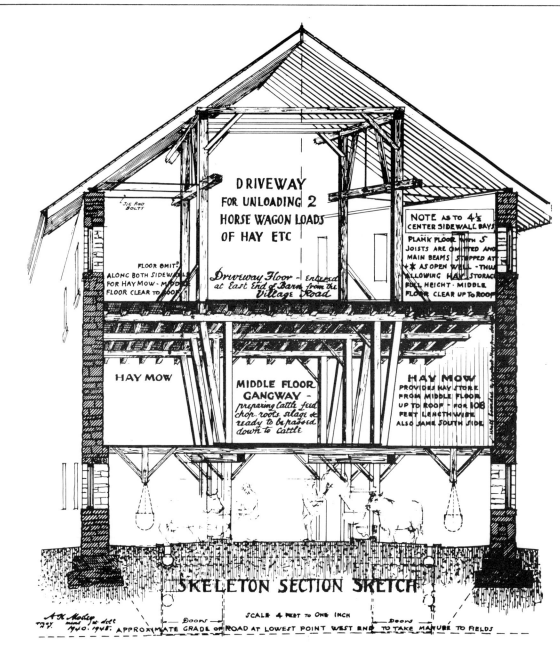

Shaker stone barn. Iron stanchions and concrete floors, noted in the detail drawing on *facing page*, were more commonly found in Shaker barns than in those of secular farmers. *Left:* Drawing inaccurately shows driveway floors extending over the mows. In fact, the mows were unobstructed all the way to the rafters.

Within the drawing:

TIE ROD BOLTS

DRIVEWAY FOR UNLOADING 2 HORSE WAGON LOADS OF HAY ETC

Driveway Floor - Entered at East End of Barn from the Village Road

FLOOR OMIT? ALONG BOTH SIDEWALL FOR HAY MOW - MYDO FLOOR CLEAR TO ROOF

NOTE AS TO 4½ CENTER SIDEWALL BAYS PLANK FLOOR WITH 5 JOISTS ARE OMITTED AND MAIN BEAMS STOPPED AT ✱ AS OPEN WELL - THUS ALLOWING HAY STORAGE FULL HEIGHT · MIDDLE FLOOR CLEAR UP TO ROOF

HAY MOW

MIDDLE FLOOR GANGWAY - *preparing cattle feed chop. roots. silage & ready to be passed down to cattle*

HAY MOW PROVIDES HAY STORE FROM MIDDLE FLOOR UP TO ROOF - FOR 108 FEET LENGTHWISE ALSO SAME SOUTH SIDE

SKELETON SECTION SKETCH

SCALE 4 FEET TO ONE INCH

Doors

Doors

1940 · 1945 · APPROXIMATE GRADE OF ROAD AT LOWEST POINT WEST END · TO TAKE MANURE TO FIELDS

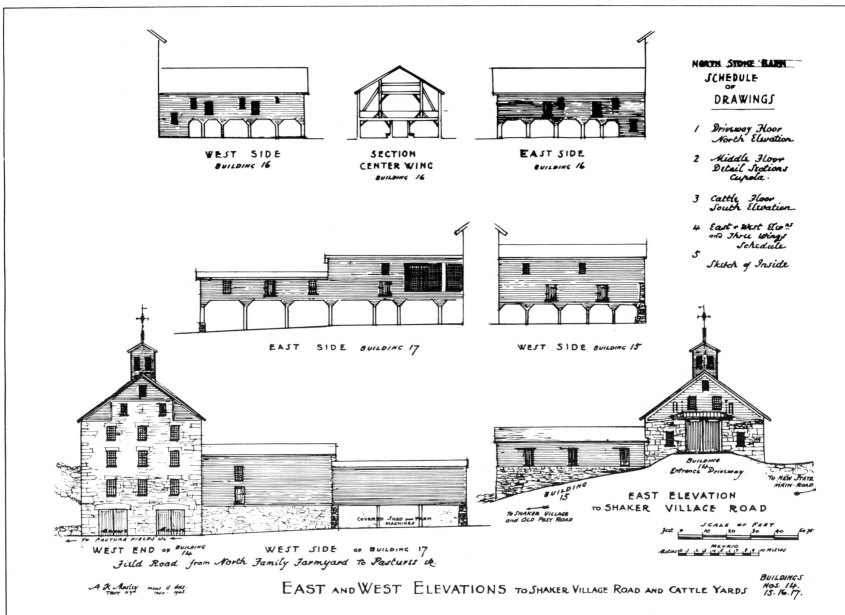

WEST SIDE
BUILDING 16

SECTION
CENTER WING
BUILDING 16

EAST SIDE
BUILDING 16

NORTH STONE BARN
SCHEDULE
OF
DRAWINGS

1 Driveway Floor
North Elevation

2 Middle Floor
Detail Sections
Cupola.

3 Cattle Floor
South Elevation

4 East & West Elev.ns
and Three Wings
Schedule

5 Sketch of Inside

EAST SIDE BUILDING 17

WEST SIDE BUILDING 15

WEST END OF BUILDING 14
Field Road from North Family Farmyard to Pastures &c.

WEST SIDE OF BUILDING 17

COVERED SHED FOR FARM MACHINES

BUILDING 15
To SHAKER VILLAGE
and OLD POST ROAD

BUILDING 14
Entrance Driveway

To NEW STATE
MAIN-ROAD

EAST ELEVATION
TO SHAKER VILLAGE ROAD

SCALE OF FEET
feet 0 10 20 30 40 50 ft.
METRIC
meters 1 2 3 4 5 6 7 8 9 10 meters

A. K. Mosley meas. & del.
TROY N.Y. 1903-1905.

EAST AND WEST ELEVATIONS TO SHAKER VILLAGE ROAD AND CATTLE YARDS

BUILDINGS
NOS. 14.
15. 16. 17.

Shaker stone barn. Wings provide additional hay storage above the open cattle shelters.

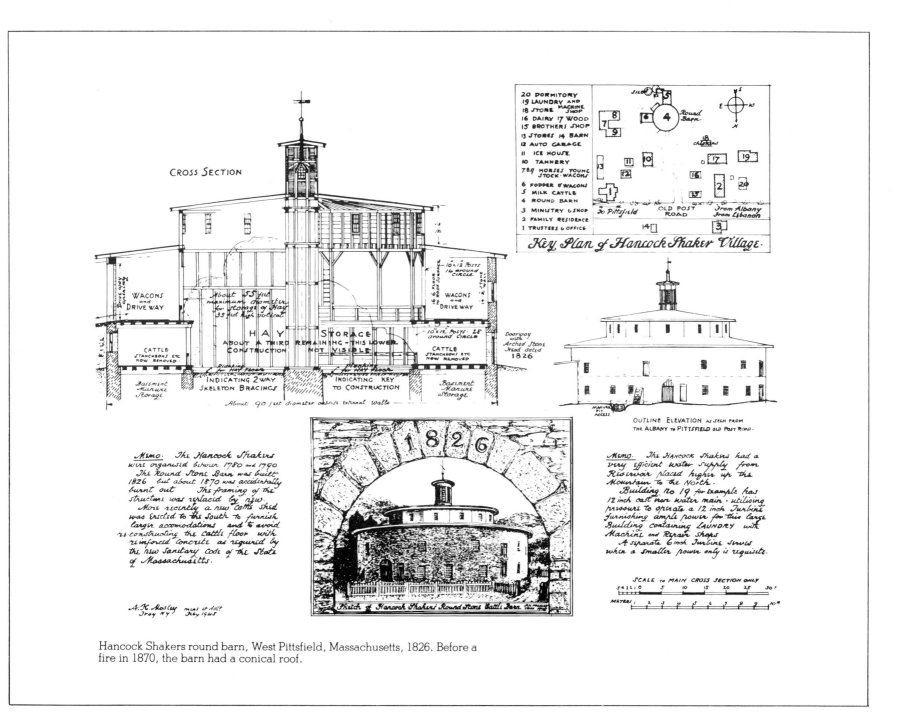

Hancock Shakers round barn, West Pittsfield, Massachusetts, 1826. Before a fire in 1870, the barn had a conical roof.

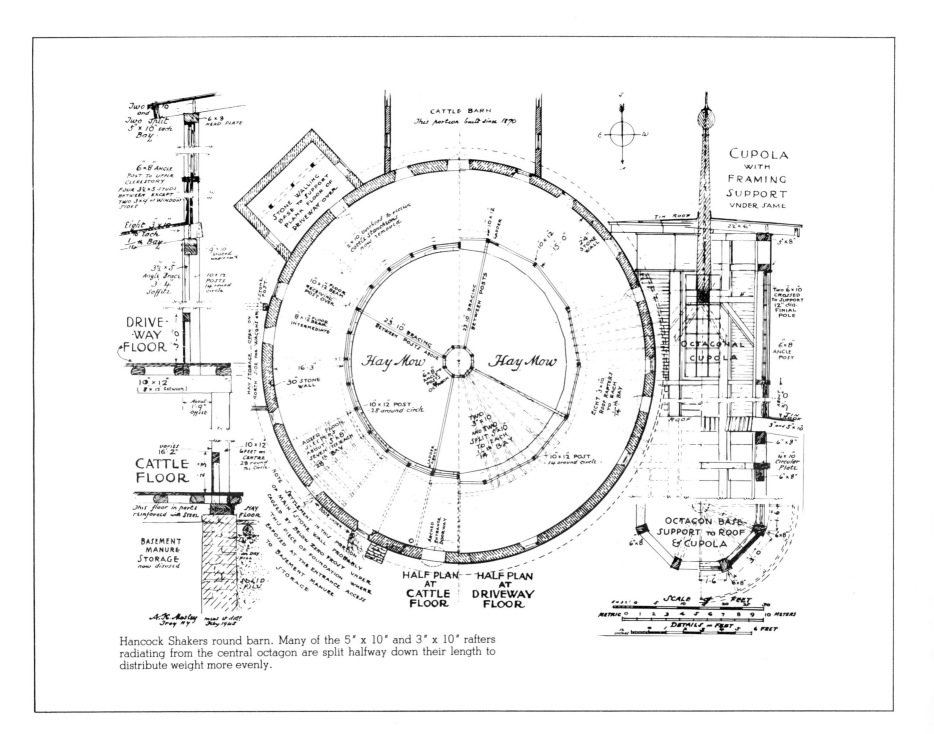

Hancock Shakers round barn. Many of the 5" x 10" and 3" x 10" rafters radiating from the central octagon are split halfway down their length to distribute weight more evenly.

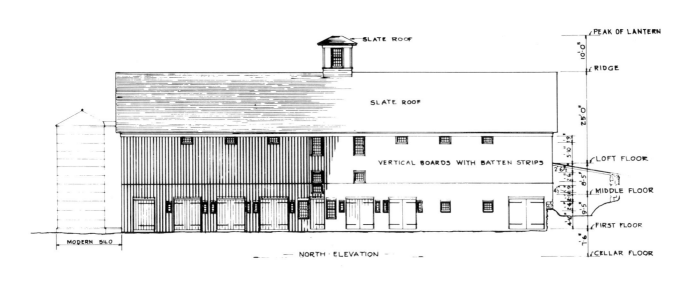

SLATE ROOF

SLATE ROOF

PEAK OF LANTERN

RIDGE

VERTICAL BOARDS WITH BATTEN STRIPS

LOFT FLOOR

MIDDLE FLOOR

FIRST FLOOR

CELLAR FLOOR

MODERN SILO

— NORTH · ELEVATION —

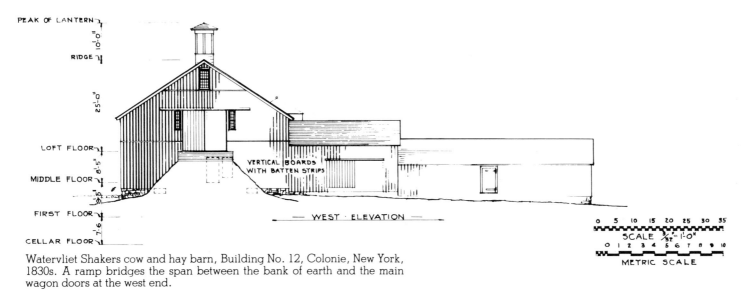

PEAK OF LANTERN

RIDGE

LOFT FLOOR

MIDDLE FLOOR

VERTICAL BOARDS WITH BATTEN STRIPS

FIRST FLOOR

CELLAR FLOOR

— WEST · ELEVATION —

SCALE 3/32" = 1'-0"

METRIC SCALE

Watervliet Shakers cow and hay barn, Building No. 12, Colonie, New York, 1830s. A ramp bridges the span between the bank of earth and the main wagon doors at the west end.

85

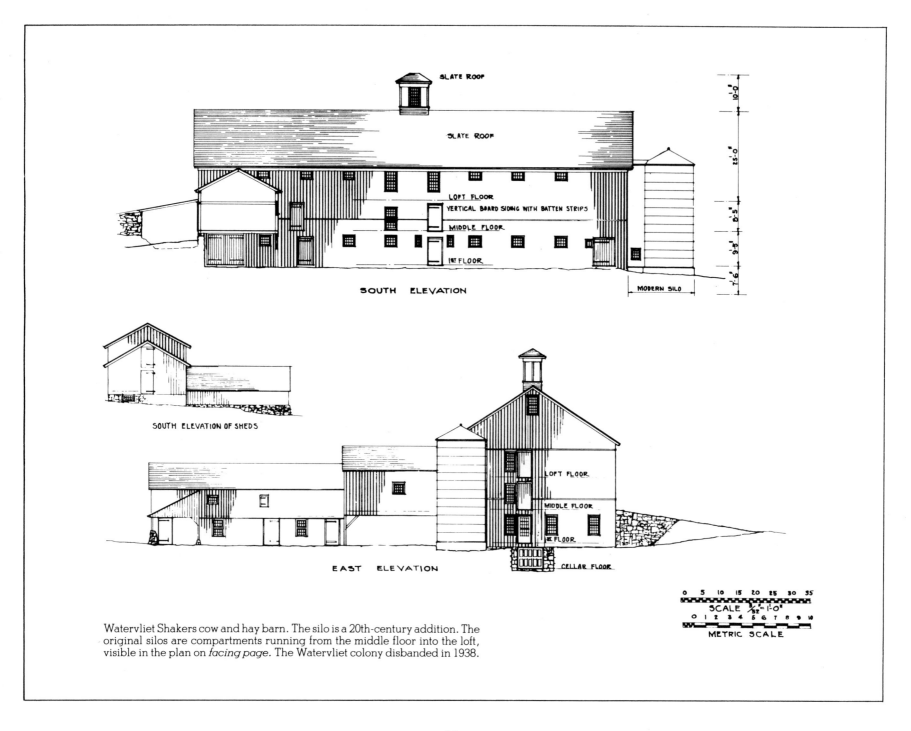

SLATE ROOF

SLATE ROOF

LOFT FLOOR

VERTICAL BOARD SIDING WITH BATTEN STRIPS

MIDDLE FLOOR

1ST FLOOR

SOUTH ELEVATION

MODERN SILO

SOUTH ELEVATION OF SHEDS

LOFT FLOOR

MIDDLE FLOOR

1ST FLOOR

EAST ELEVATION

CELLAR FLOOR

0 5 10 15 20 25 30 35

SCALE 3/32" = 1'-0"

0 1 2 3 4 5 6 7 8 9 10

METRIC SCALE

Watervliet Shakers cow and hay barn. The silo is a 20th-century addition. The original silos are compartments running from the middle floor into the loft, visible in the plan on *facing page.* The Watervliet colony disbanded in 1938.

86

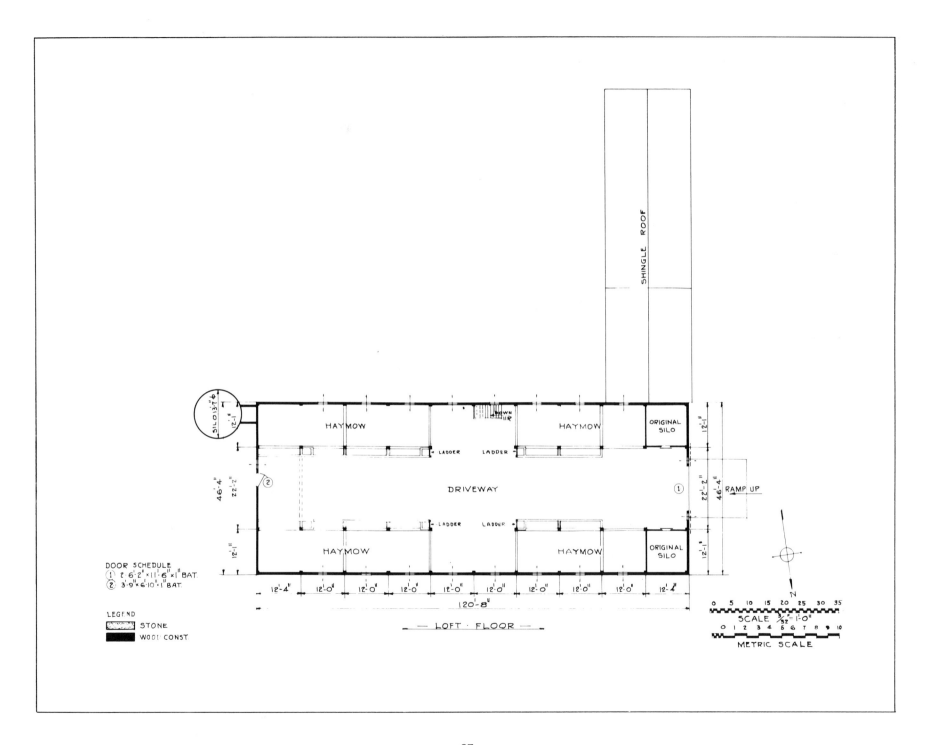

SHINGLE ROOF

SILO 13'-7"∅

HAYMOW

DOWN 11R

HAYMOW

ORIGINAL SILO

12'-1"

LADDER   LADDER

46'-4"

22'-2"

2

DRIVEWAY

1

RAMP UP

22'-2"

46'-4"

12'-1"

LADDER   LADDER

12'-1"

HAYMOW

HAYMOW

ORIGINAL SILO

12'-1"

N

12'-4" 12'-0" 12'-0" 12'-0" 12'-0" 12'-0" 12'-0" 12'-0" 12'-0" 12'-4"

120'-8"

— LOFT · FLOOR —

DOOR SCHEDULE
1 2'-6'-2" × 11'-6" × 1" BAT.
2 3'-9" × 6'-10" × 1" BAT.

LEGEND
▨ STONE
█ WOOD CONST.

0 5 10 15 20 25 30 35
SCALE 3/32" = 1'-0"
0 1 2 3 4 5 6 7 8 9 10
METRIC SCALE

87

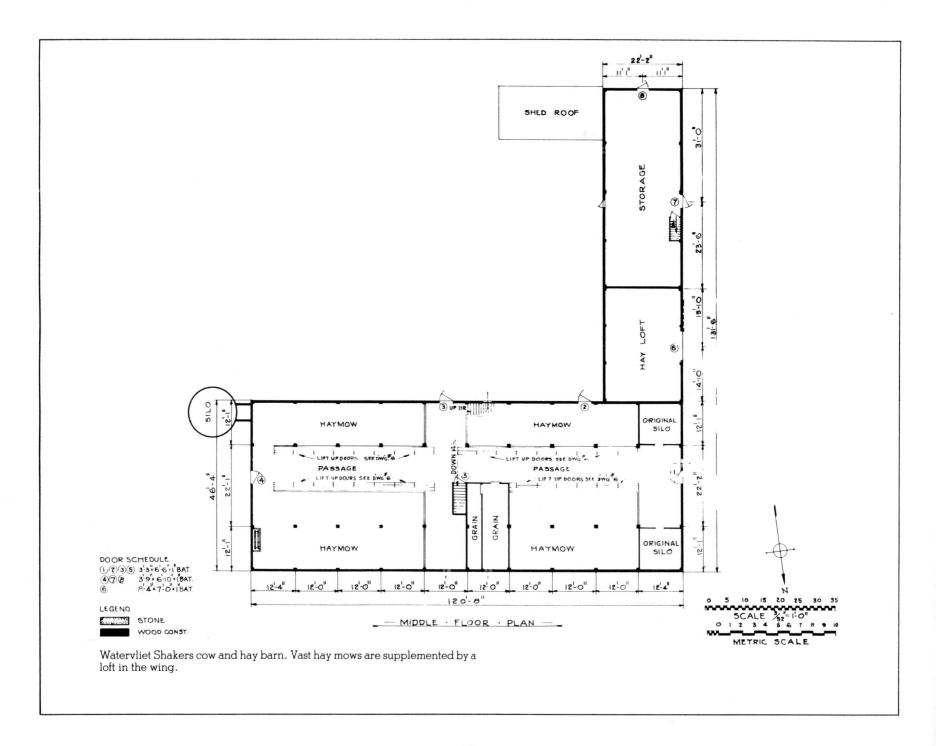

SHED ROOF

STORAGE

HAY LOFT

SILO

HAYMOW

HAYMOW

HAYMOW

HAYMOW

ORIGINAL SILO

ORIGINAL SILO

PASSAGE

PASSAGE

GRAIN

GRAIN

LIFT UP DOORS SEE DWG# 6

LIFT UP DOORS SEE DWG# 6

LIFT UP DOORS SEE DWG# 6

LIFT UP DOORS SEE DWG# 6

UP 11R

DOWN 14

DOOR SCHEDULE
①②③⑤ 3'-3" 6'-6" × 1 BAT.
④⑦⑧ 3'-9" × 6'-10" × 1 BAT.
⑥ 8'-4" × 7'-0" × 1 BAT.

LEGEND
STONE
WOOD CONST.

— MIDDLE · FLOOR · PLAN —

0 5 10 15 20 25 30 35
SCALE 3/32" = 1'-0"
0 1 2 3 4 5 6 7 8 9 10
METRIC SCALE

N

Watervliet Shakers cow and hay barn. Vast hay mows are supplemented by a loft in the wing.

22'-2"
11'-1" 11'-1"
31'-0"
23'-6"
15'-0"
13'-6"
14'-0"
12'-1"
22'-2"
12'-1"
46'-4"
22'-2"
12'-1"
12'-4" 12'-0" 12'-0" 12'-0" 12'-0" 12'-0" 12'-0" 12'-0" 12'-0" 12'-4"
120'-8"

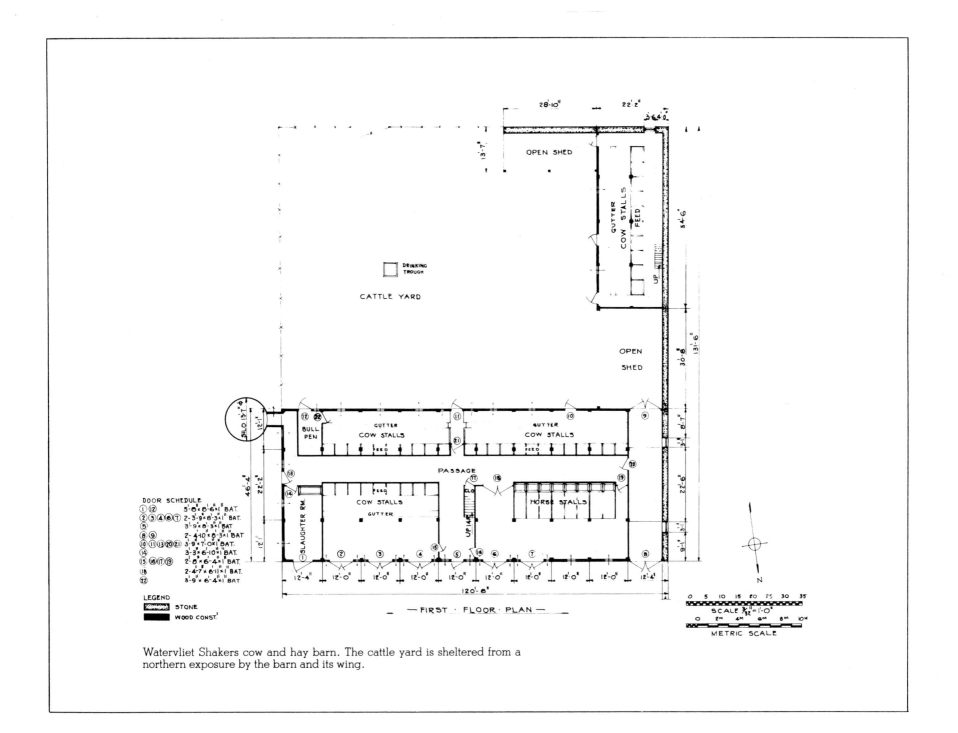

28'-10" 22'-2"

13'-7"

OPEN SHED

GUTTER COW STALLS FEED

54'-6"

UP

DRINKING
TROUGH

CATTLE YARD

OPEN

SHED

30'-8" 131'-6"

SILO 13'-7" ∅

12'-1"

BULL
PEN

GUTTER
COW STALLS FEED

GUTTER
COW STALLS FEED

8'-7"

46'-4" 22'-2"

13

14

SLAUGHTER RM.

PASSAGE

CLO.

UP

COW STALLS

GUTTER FEED

HORSE STALLS

12'-0"

DOOR SCHEDULE
①⑫ 5'-0"x8'-6"x1" BAT.
②③⑥⑦ 2-3'-9"x8'-3"x1" BAT.
⑤ 3'-9"x8-3"x1" BAT.
⑧⑨ 2-4'-10"x8'-3"x1" BAT.
⑩⑪⑬⑳㉑ 3'-9"x7'-0"x1" BAT.
⑭ 3'-3"x6'-10"x1" BAT.
⑮⑯⑰⑲ 2'-6"x6-4"x1" BAT.
⑱ 2-4'-7"x6'-11"x1" BAT.
㉒ 3'-9"x6-4"x1" BAT.

LEGEND
▨ STONE
▉ WOOD CONST.

12'-4" 12'-0" 12'-0" 12'-0" 12'-0" 12'-0" 12'-0" 12'-0" 12'-0" 12'-4"

120'-8"

— FIRST · FLOOR · PLAN —

N

SCALE ³⁄₁₆"=1'-0"
0 5 10 15 20 25 30 35

0 2ᴹ 4ᴹ 6ᴹ 8ᴹ 10ᴹ
METRIC SCALE

Watervliet Shakers cow and hay barn. The cattle yard is sheltered from a
northern exposure by the barn and its wing.

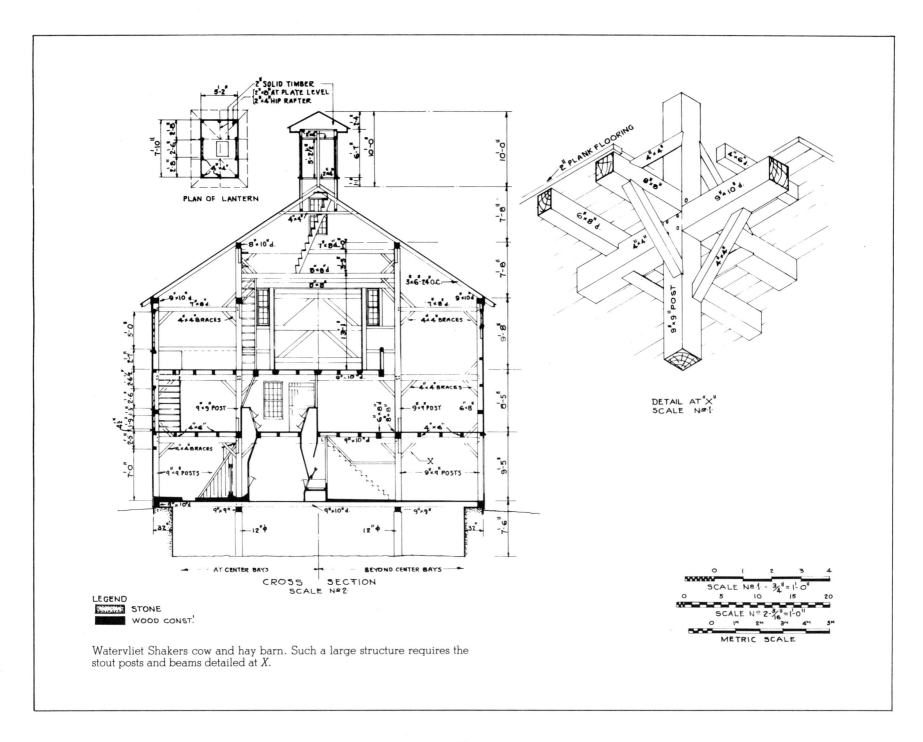

PLAN OF LANTERN

2" SOLID TIMBER
(2"×8" AT PLATE LEVEL
(2"×4" HIP RAFTER

DETAIL AT "X"
SCALE Nº1

2" PLANK FLOORING

9"×9" POST

CROSS SECTION
SCALE Nº 2

AT CENTER BAYS   BEYOND CENTER BAYS

LEGEND

STONE

WOOD CONST.'

Watervliet Shakers cow and hay barn. Such a large structure requires the stout posts and beams detailed at *X*.

SCALE Nº1 - 3/4" = 1'-0"

SCALE Nº 2 - 3/16" = 1'-0"

METRIC SCALE

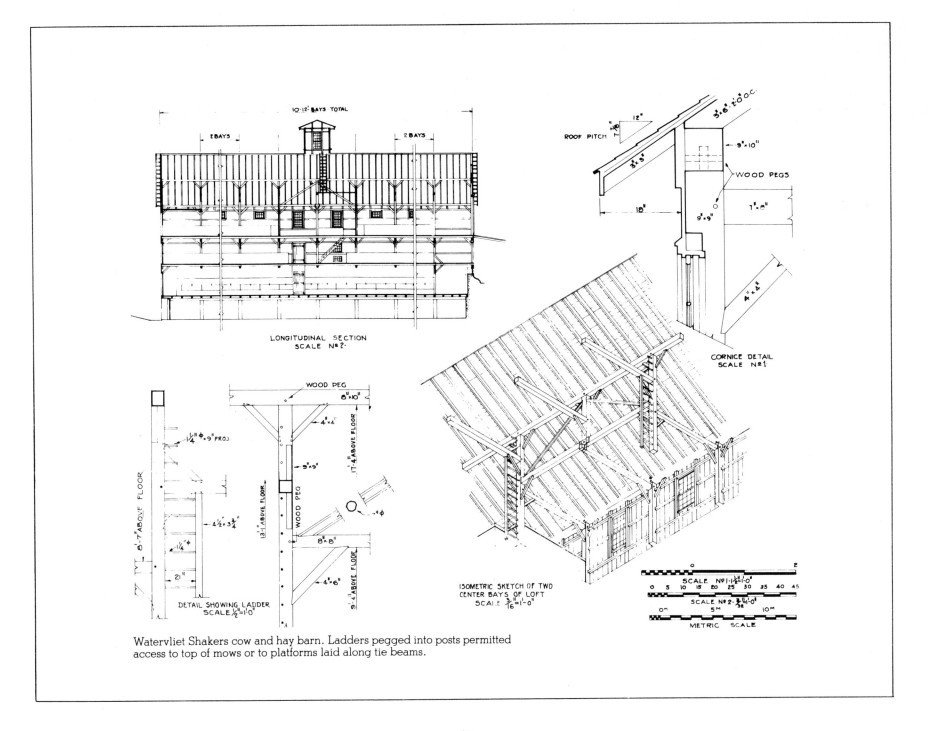

LONGITUDINAL SECTION
SCALE № 2

CORNICE DETAIL
SCALE № 1

DETAIL SHOWING LADDER
SCALE ½"=1'-0"

ISOMETRIC SKETCH OF TWO
CENTER BAYS OF LOFT
SCALE 3/16"=1'-0"

SCALE № 1-1½"=1'-0"
SCALE № 2-3/32"=1'-0"
METRIC SCALE

Watervliet Shakers cow and hay barn. Ladders pegged into posts permitted
access to top of mows or to platforms laid along tie beams.

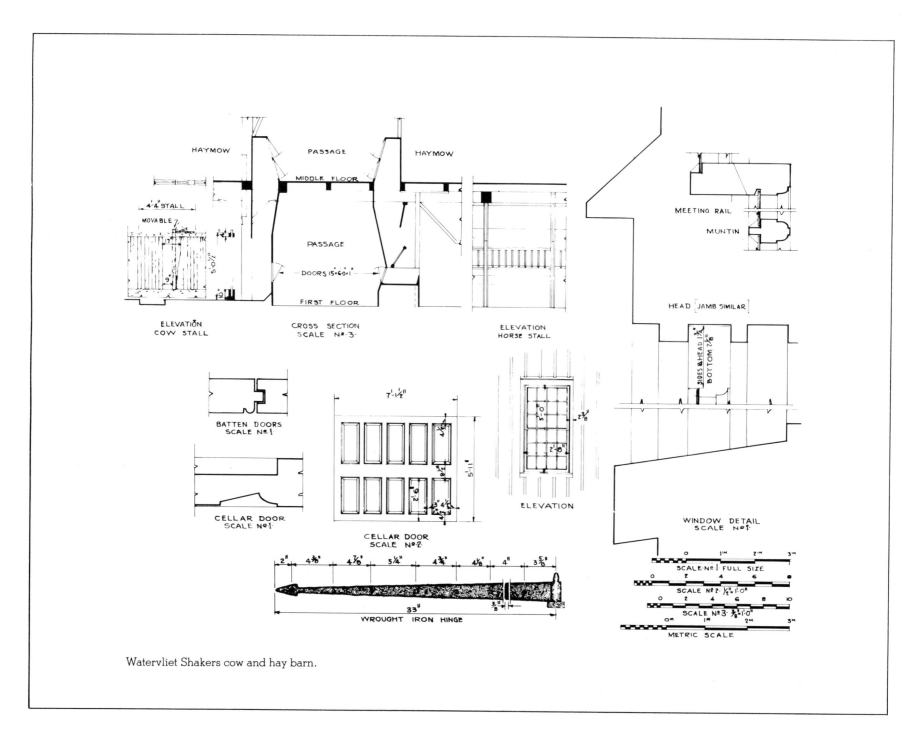

HAYMOW PASSAGE HAYMOW

MIDDLE FLOOR

4'-4" STALL

MOVABLE

PASSAGE

DOORS 15·60·1

FIRST FLOOR

ELEVATION
COW STALL

CROSS SECTION
SCALE N⁰ 3·

ELEVATION
HORSE STALL

MEETING RAIL

MUNTIN

HEAD [JAMB SIMILAR]

BATTEN DOORS
SCALE N⁰ 1·

CELLAR DOOR
SCALE N⁰ 1·

7'-1½"

5'-11"

CELLAR DOOR
SCALE N⁰ 2·

ELEVATION

SIDES & HEAD 1¾"
BOTTOM 2⅛"

WINDOW DETAIL
SCALE N⁰ 1·

2" 4⅜" 4⅞" 5¼" 4¾" 4⅛" 4" 3⅝"

33"

⅜"

WROUGHT IRON HINGE

SCALE N⁰ 1 FULL SIZE

SCALE N⁰ 2· ½"=1'·0"

SCALE N⁰ 3· ⅜"=1'·0"

METRIC SCALE

Watervliet Shakers cow and hay barn.

92

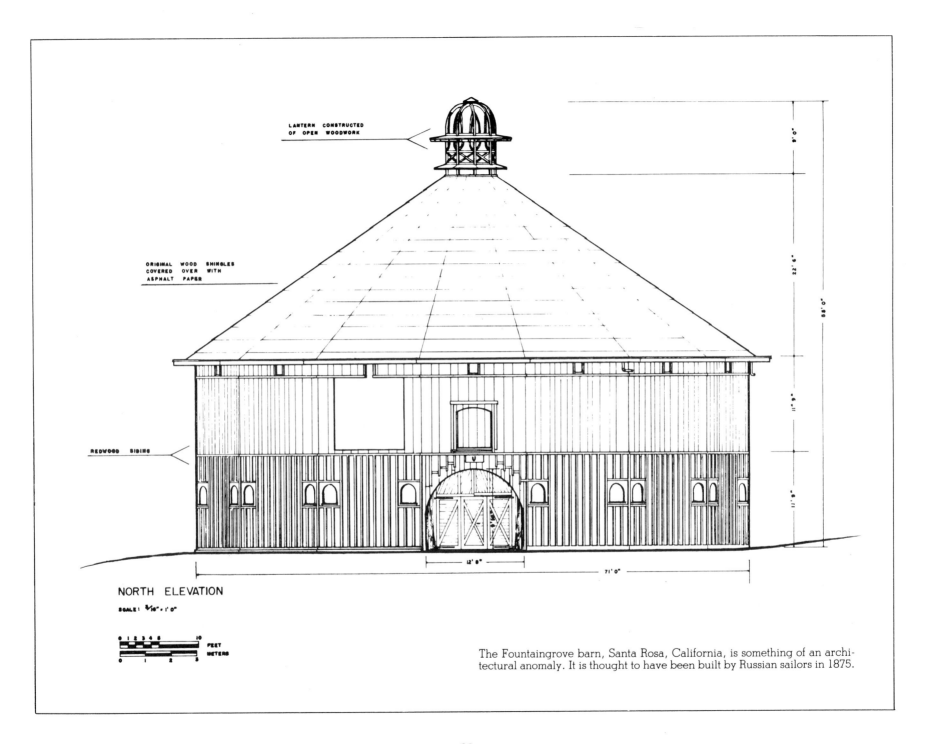

LANTERN CONSTRUCTED
OF OPEN WOODWORK

ORIGINAL WOOD SHINGLES
COVERED OVER WITH
ASPHALT PAPER

REDWOOD SIDING

9' 0"

22' 6"

53' 0"

11" 9"

11" 9"

12' 8"

71' 0"

NORTH ELEVATION

SCALE: 3/16" = 1' 0"

0 1 2 3 4 5   10   FEET
   0   1   2   5   METERS

The Fountaingrove barn, Santa Rosa, California, is something of an architectural anomaly. It is thought to have been built by Russian sailors in 1875.

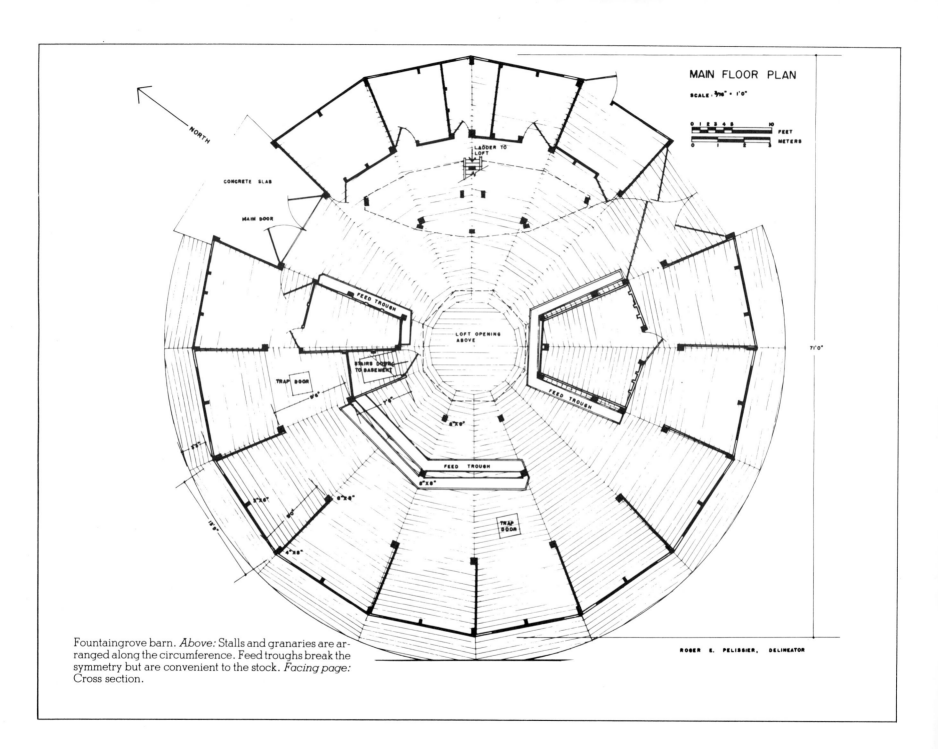

MAIN FLOOR PLAN

SCALE: 3/16" = 1'0"

NORTH

CONCRETE SLAB

MAIN DOOR

LADDER TO LOFT

FEED TROUGH

LOFT OPENING ABOVE

STAIRS DOWN TO BASEMENT

TRAP DOOR

FEED TROUGH

FEED TROUGH

TRAP DOOR

71'0"

ROGER E. PELISSIER, DELINEATOR

Fountaingrove barn. *Above:* Stalls and granaries are arranged along the circumference. Feed troughs break the symmetry but are convenient to the stock. *Facing page:* Cross section.

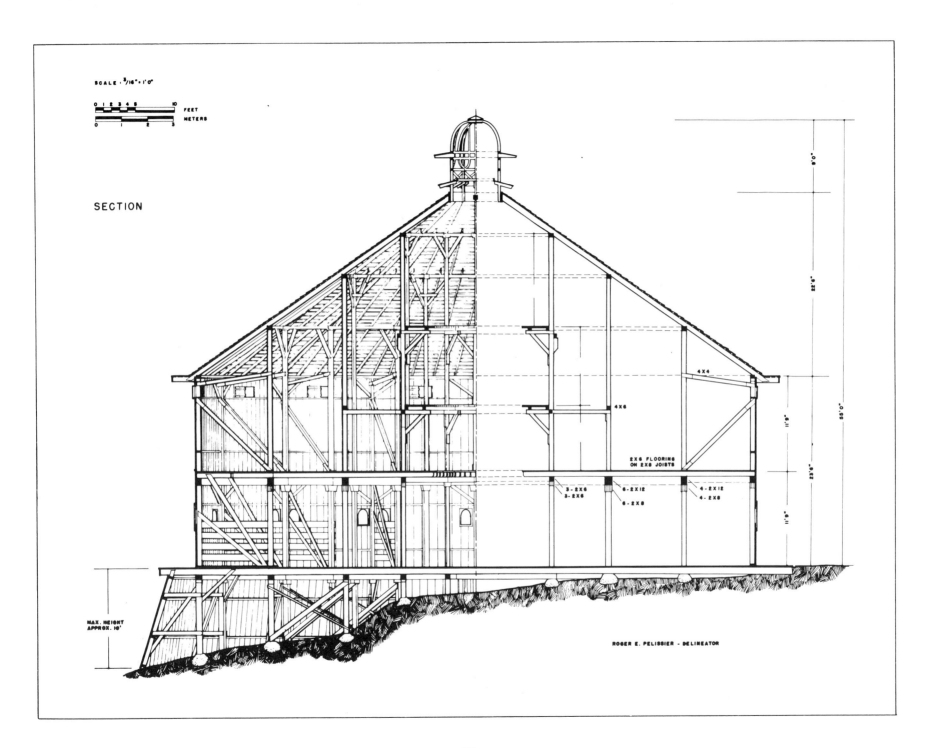

SECTION

SCALE · 3/16"=1'0"

0 1 2 3 4 5   10   FEET
              METERS
0   1   2   3

4X4

4X6

2X6 FLOORING
ON 2X8 JOISTS

3-2X6
3-2X6

6-2X12
6-2X8

4-2X12
4-2X8

9'0"

22'6"

55'0"

11'9"

23'6"

11'9"

MAX. HEIGHT
APPROX. 19'

ROGER E. PELISSIER - DELINEATOR

Left: An octagonal barn, Washington County, Pennsylvania, 1885. Four center posts (at *a* in the floor plans, *facing page, top*), support the internal frame. They are a foot square up to the second floor, where there is an offset of 2", visible in cross section, *facing page, bottom left,* to receive the sills and plates. *Facing page, bottom right:* Iron straps and bolts fasten lapped sills and plates.

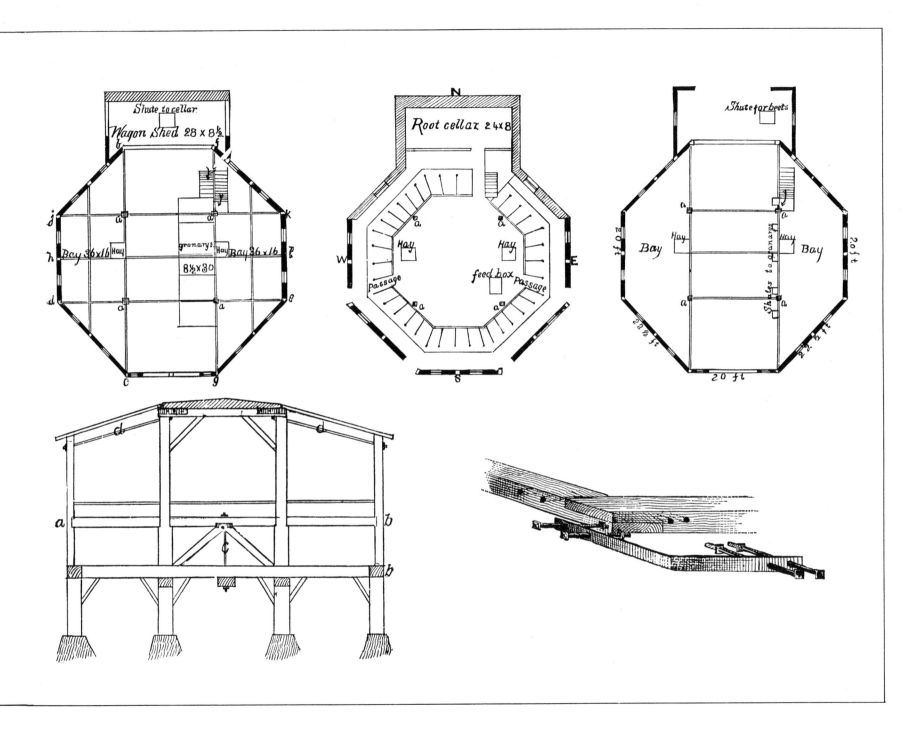

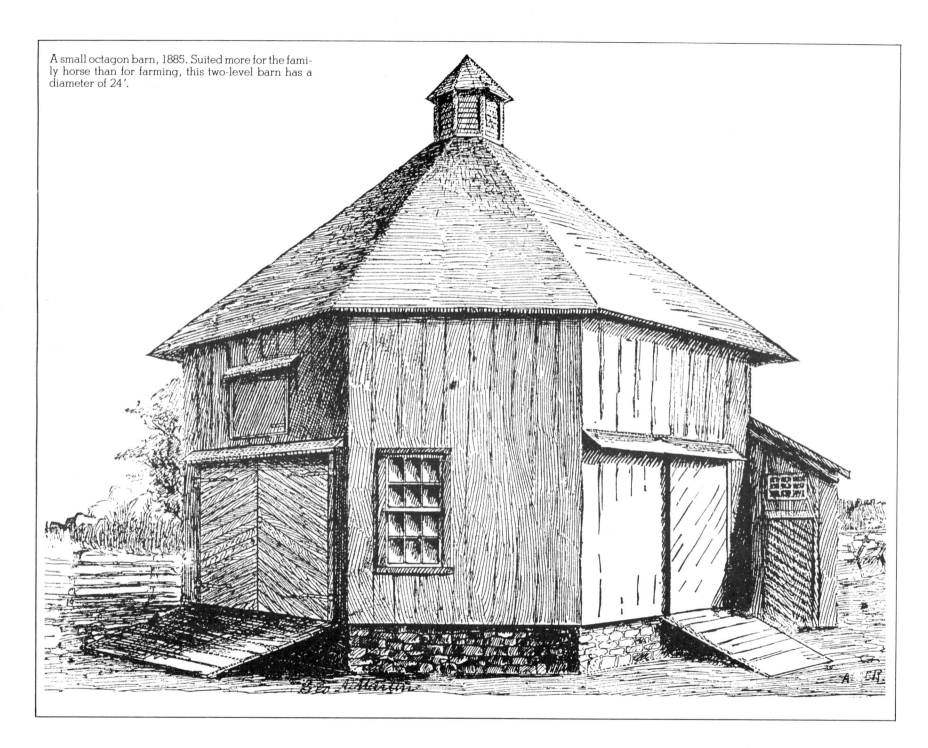

A small octagon barn, 1885. Suited more for the family horse than for farming, this two-level barn has a diameter of 24'.

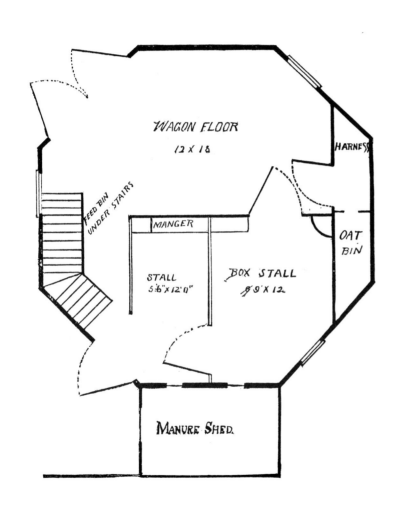

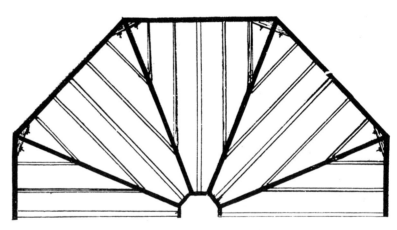

The first floor, *left*, is divided in the center by a partition, upon one side of which are a 9' x 12' box stall, an open stall, an oat bin, and a passageway. The other half is occupied by a carriage floor and harness room. Hay is supplied through the ventilator shaft from the loft. The main rafters, *above*, are 2" x 6" x 16½' and extend from each angle of the plate to the ventilator frame, and one 15½' long from the center of each plate. The shorter members are 2" x 4" x 12' and 2" x 4" x 7'. A block is bolted to each inner angle of the plate to prevent spreading.

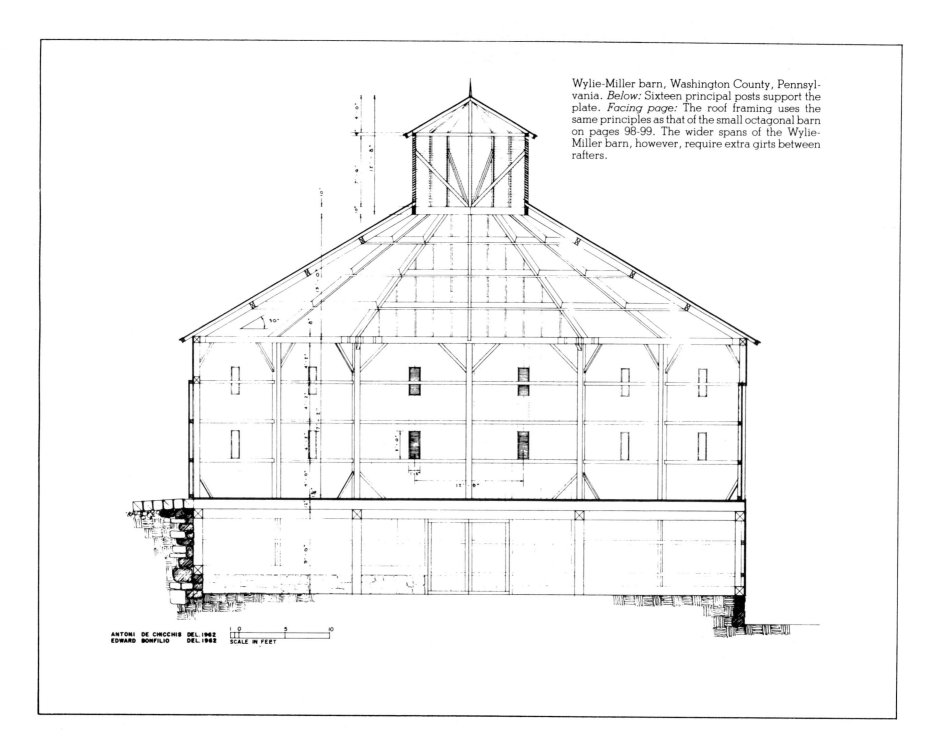

Wylie-Miller barn, Washington County, Pennsylvania. *Below:* Sixteen principal posts support the plate. *Facing page:* The roof framing uses the same principles as that of the small octagonal barn on pages 98-99. The wider spans of the Wylie-Miller barn, however, require extra girts between rafters.

ANTONI DE CHICCHIS DEL. 1962
EDWARD BONFILIO DEL. 1962
SCALE IN FEET

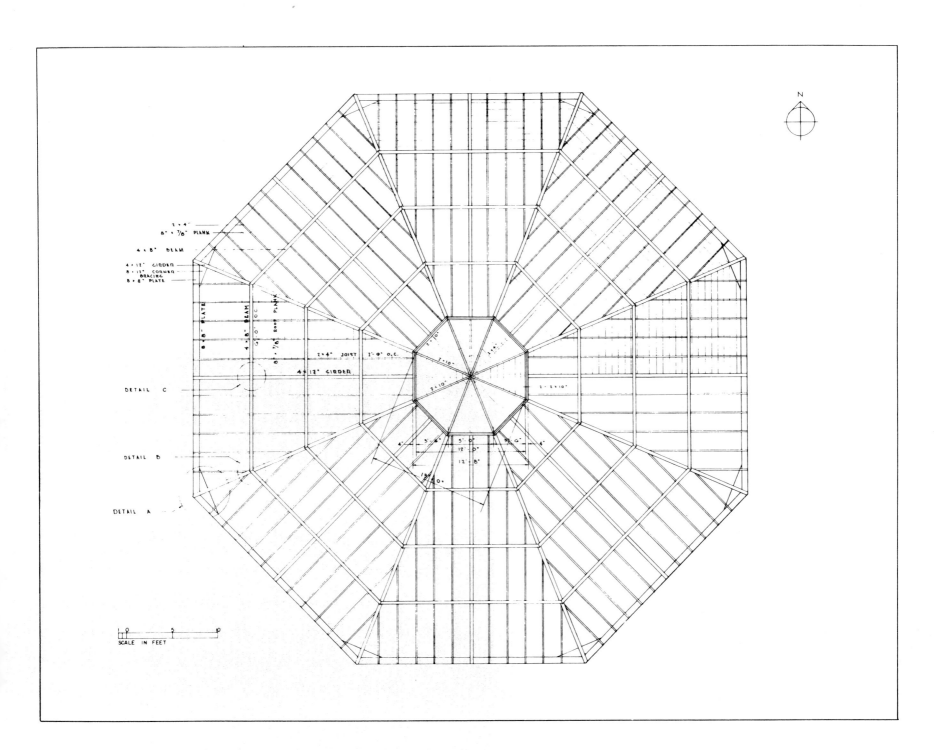

2 x 4"

8" x 7/8" PLANK

4 x 8" BEAM

4 x 12" GIRDER
8 x 12" CORNER
BRACING
8 x 8" PLATE

8 x 8" PLATE

4 x 8" BEAM

8 x 7/8" ROOF PLANK

2 x 4" JOIST 2'-0" O.C.

4 x 12" GIRDER

DETAIL C

DETAIL B

DETAIL A

1'-10"

1'-10"

1'-10"

1'-10"

1 - 2x10"

4" 5'-6" 5'-0" 3'-6" 4"

12'-0"

12'-8"

13'-0"

N

SCALE IN FEET

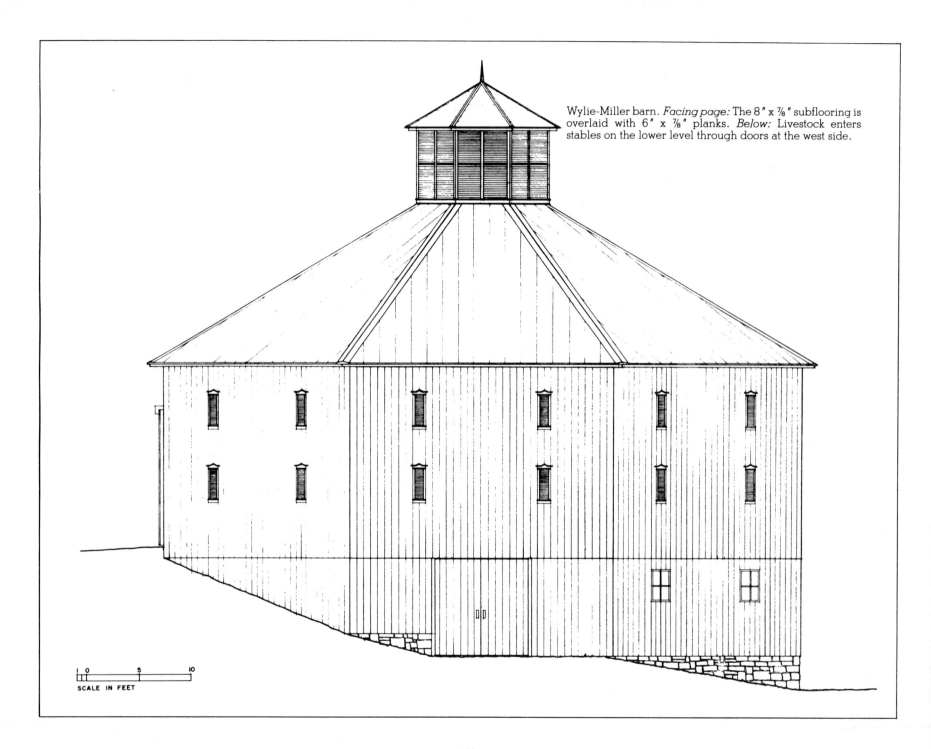

Wylie-Miller barn. *Facing page:* The 8" x ⅞" subflooring is overlaid with 6" x ⅞" planks. *Below:* Livestock enters stables on the lower level through doors at the west side.

I 0   5   10

SCALE IN FEET

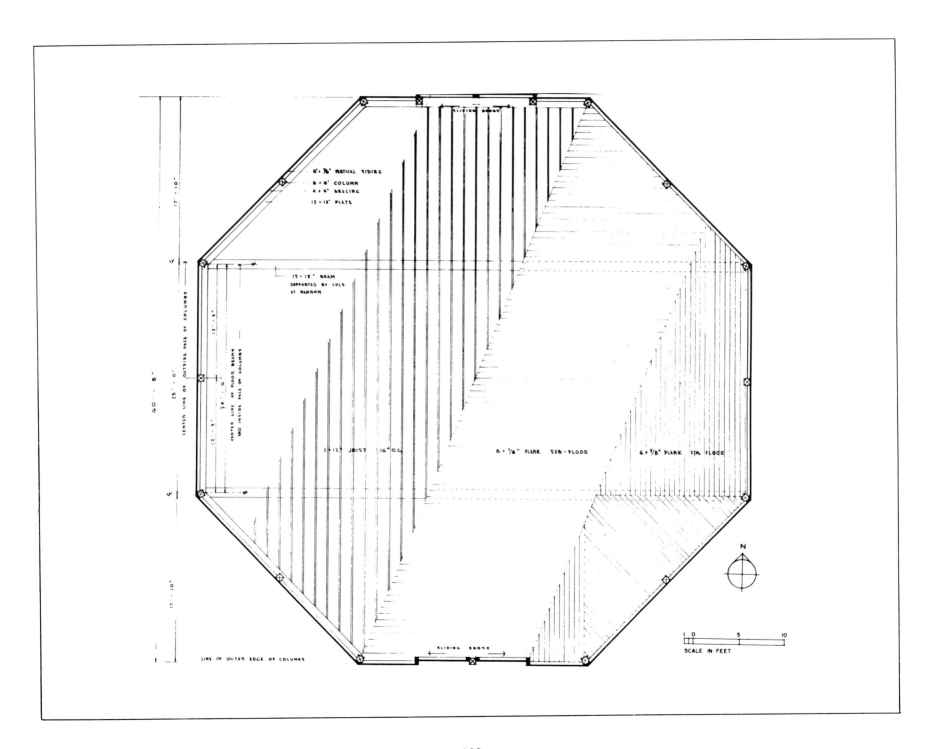

8 x ⅞" VERTICAL SIDING
8 x 8" COLUMN
4 x 4" BRACING
12 x 12" PLATE

12 x 12" BEAM
SUPPORTED BY COLS.
AT RANDOM

CENTER LINE OF OUTSIDE FACE OF COLUMNS

CENTER LINE OF FLOOR BEAMS
AND INSIDE FACE OF COLUMNS

12'-3"   24'-6"   12'-3"

25'-0"

60'-8"

17'-10"   6'   6'   17'-10"

2 x 12" JOIST 16" O.C.

8 x ⅞" PLANK SUB-FLOOR

6 x ⅞" PLANK FIN FLOOR

SLIDING DOORS

SLIDING DOORS

LINE OF OUTER EDGE OF COLUMNS

N

SCALE IN FEET
1 0   5   10

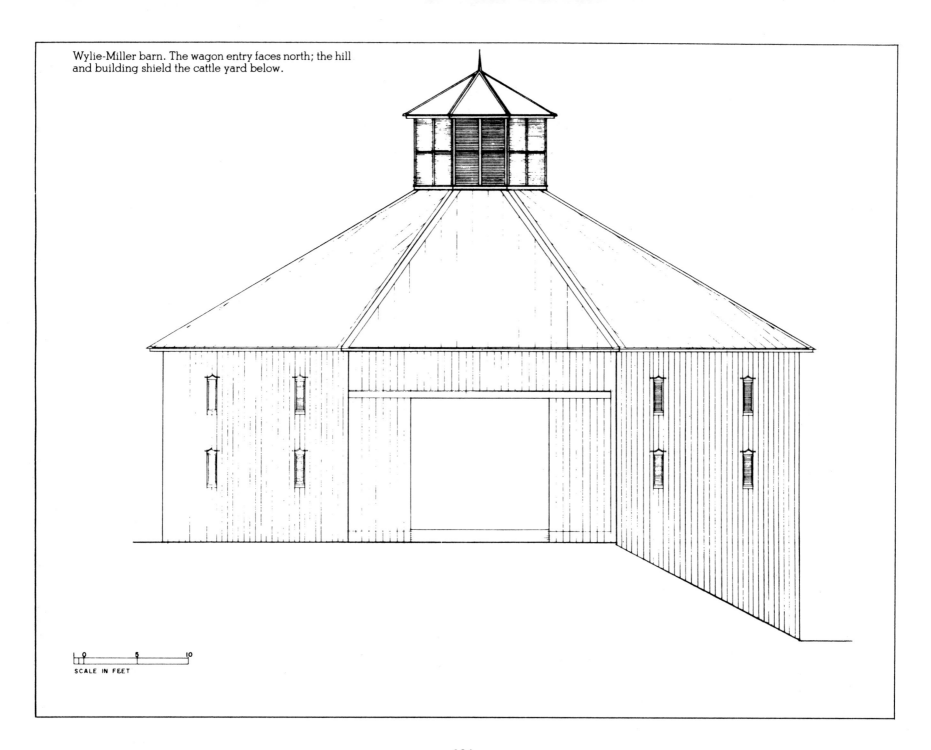

Wylie-Miller barn. The wagon entry faces north; the hill and building shield the cattle yard below.

SCALE IN FEET

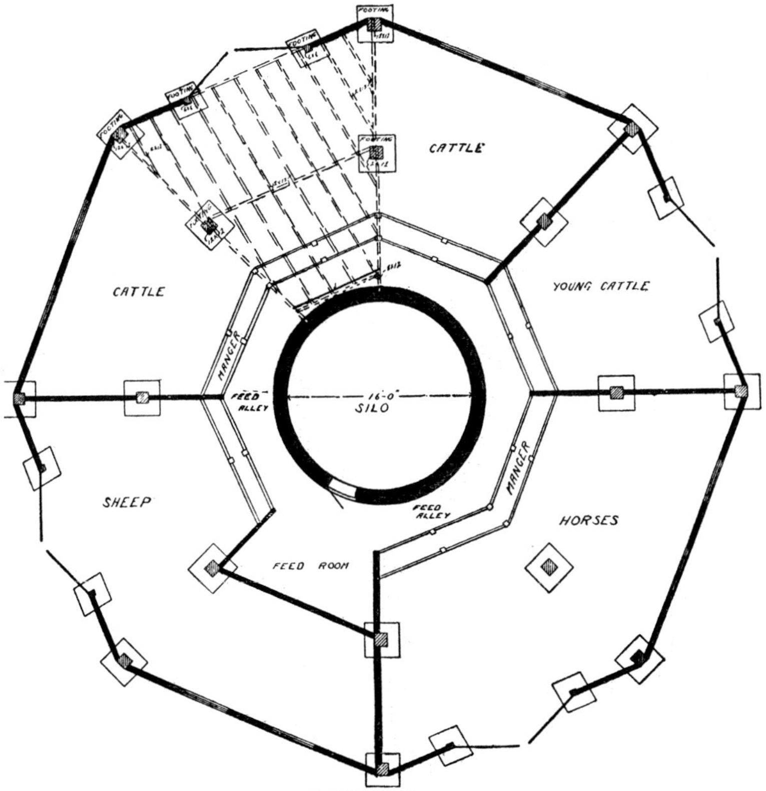
CATTLE
CATTLE
YOUNG CATTLE
MANGER
FEED ALLEY
16'-0"
SILO
MANGER
FEED ALLEY
SHEEP
HORSES
FEED ROOM

FLOOR PLAN

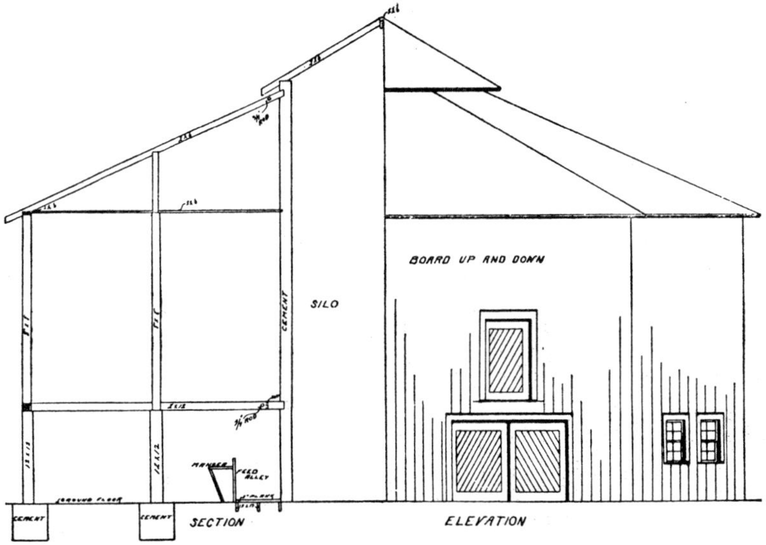
BOARD UP AND DOWN
SILO
CEMENT
CEMENT
CEMENT
SECTION
ELEVATION

The modern (1908) octagonal cattle barn, *above,* is built around a 32'-high concrete silo. The frame is tied into the silo and takes advantage of its stability. *Right:* The location of the feed alley and mangers reduces the work of feeding to a minimum. The 12" x 12" posts are sunk in concrete footings.

A prize-winning design, illustrated on pages 116-117.

victorian barns

As far as the Shakers went with their functional aesthetic, some builders of barns in the second half of the 19th century went at least as far in the opposite direction. This is not to say that the functions of a barn were inadequately provided for; on the contrary, most Victorian barns reflected "the progress of the age" in sanitation, plumbing, ventilation, building materials, and in their provisions for the ever-improving farm machinery. What distinguished Victorian builders was that, in addition to all these practical and necessary features, ornamentation was applied to doors, windows, and roof lines in an effort to "spruce up" the increasingly prosperous farms of North America. These accents were not strictly gratuitous, however, as elaboration followed form in most instances. The effect on smaller barns was to invest them with a cottage-like quaintness, and the effect on the larger ones was to enliven what otherwise might be hulking monoliths.

Less floor space was now being given over to the storage of hay and other crops compared with that devoted to livestock in many of the larger barns.

Seen for the first time are sprawling wings extending from a central cube, a design aimed at accommodating cows, sheep, horses, oxen, hogs, and poultry as well as machinery and tools under one roof and above ground. Such a configuration anticipates the long, narrow barns of the early 20th century. Only the most successful farmers could possibly afford such an ambitious structure, regardless of its economy of operation and maintenance. A man of humbler vision had to satisfy himself with lesser rewards.

In addition to the fancywork he might choose to apply to his barn, a farmer might order new steel-track sliding doors, or install sash windows which could be purchased for about four dollars each in 1880. Previously, barn windows were usually either board & batten hatches or fixed glass panes. Besides their obvious function, moveable glass windows gave a barn a more finished, proper look at a time when both utility and style were increasingly important in some of the well-established rural areas.

In 1866, *The American Agriculturist* sponsored a barn design competition aimed at encouraging the building of well-planned, efficient and, above all, economical multiple-purpose barns. The third-place winner, *above*, designed by an architect from Massachusetts, is evidence that the price of economy can be dear. The imposing Gothic facade might be more appropriately sited on the grounds of a manicured estate. The third-place barn is pictured on pages 108-111.

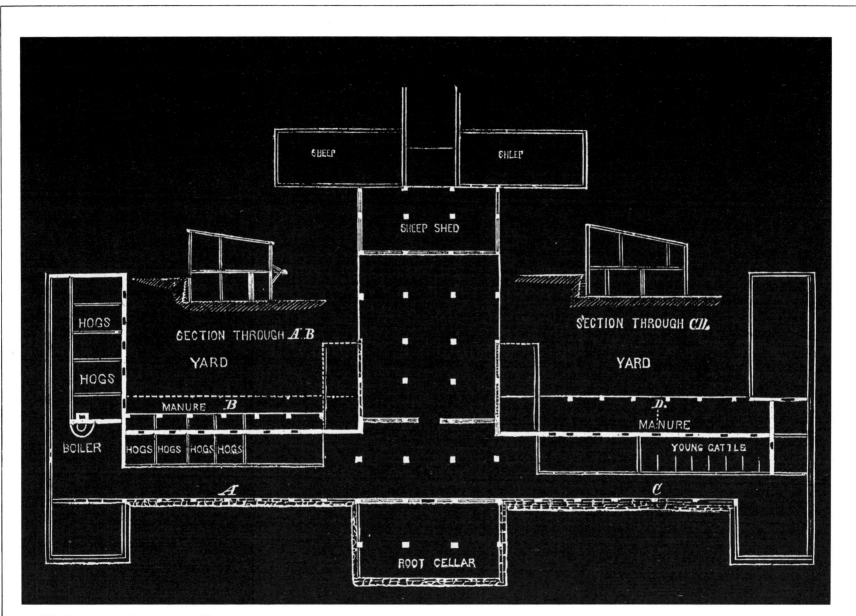

The large, central structure is filled with hay and feed and is intentionally isolated from the "unhealthy exhalations" of the livestock, a problem not even good ventilation (by the standards of 1866) could overcome. The wings house hogs, manure, and young cattle.

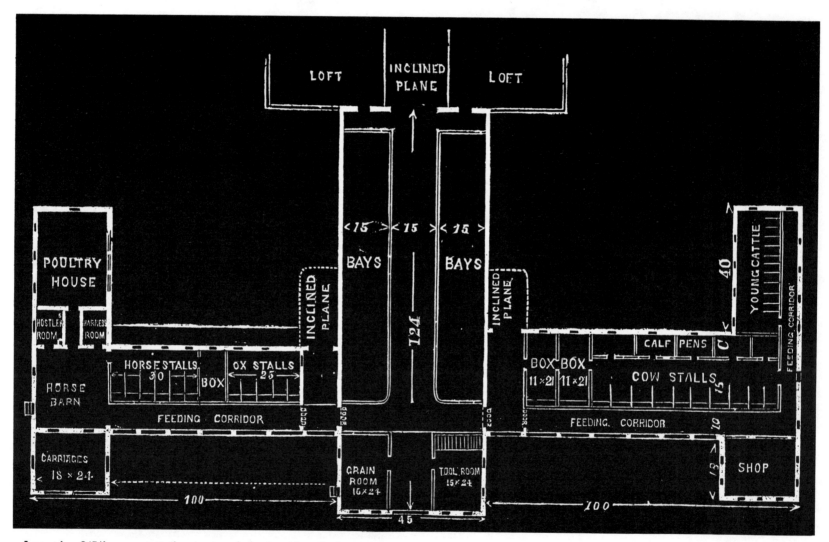

A corridor, 245' long, permits the running of a feed cart from the horses at one
end to the cows at the other. Animals can file out to the yard through a door at
each side of the main barn.

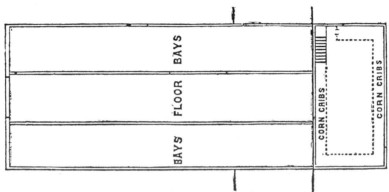

A ramp runs up to the main floor of the main barn. Each bay measures approximately 80' x 15' x 24'; combined, they hold 100 tons of hay.

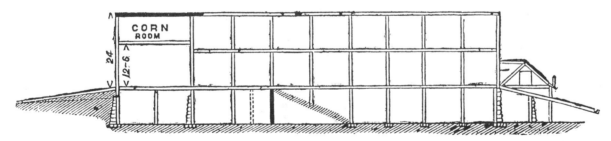

The section through the main barn reveals only the major framing elements of what would be a very expensive, elaborate barn, regardless of its external embellishments.

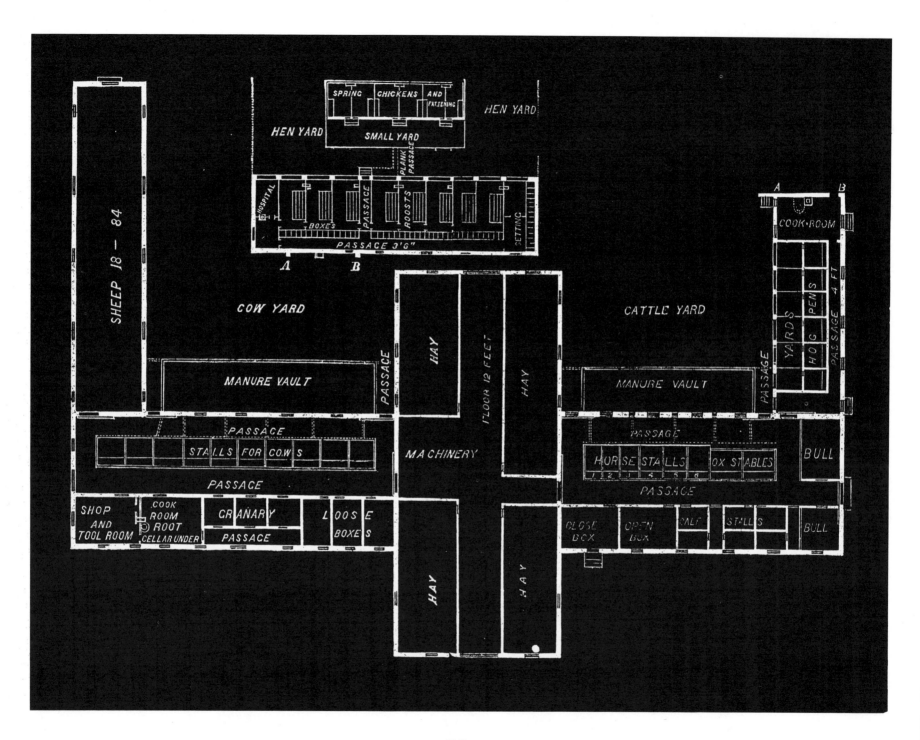

A more subdued, but equally-complete barn, all on one level, *above and facing page,* garnered the $100 second prize in the *Agriculturist's* competition. Its spare geometry includes a spired cupola and a clipped gable, or "jerkinhead" roof on the main barn.

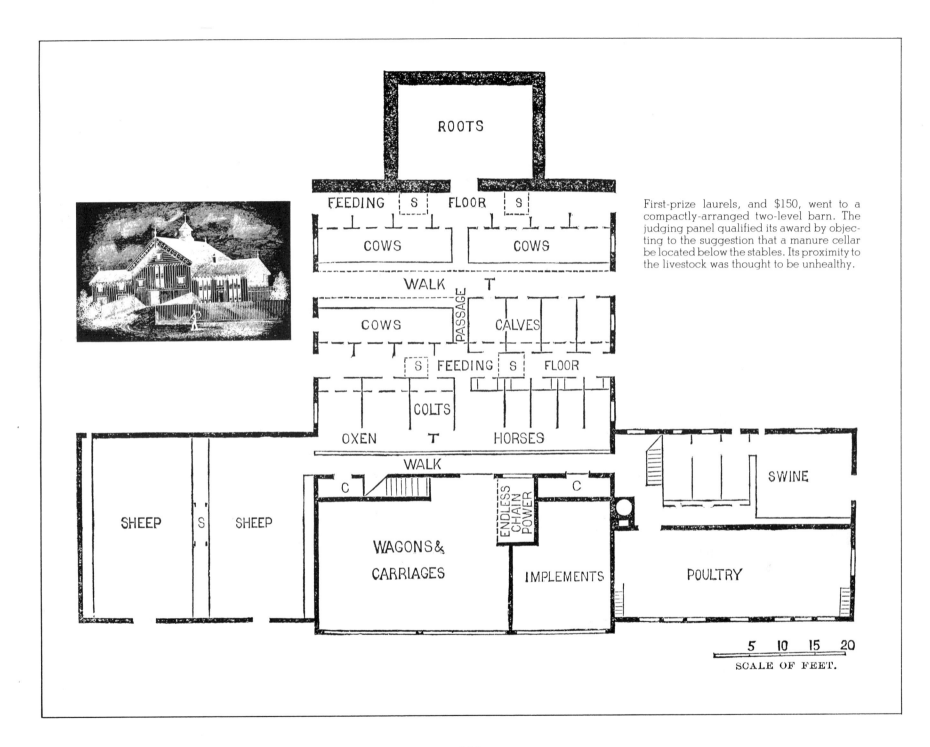

ROOTS

FEEDING S FLOOR S

COWS   COWS

WALK   T

COWS   PASSAGE   CALVES

S FEEDING S FLOOR

COLTS

OXEN   T   HORSES

WALK

C   C   SWINE

ENDLESS CHAIN POWER

SHEEP S SHEEP

WAGONS & CARRIAGES   IMPLEMENTS   POULTRY

First-prize laurels, and $150, went to a compactly-arranged two-level barn. The judging panel qualified its award by objecting to the suggestion that a manure cellar be located below the stables. Its proximity to the livestock was thought to be unhealthy.

5   10   15   20
SCALE OF FEET.

114

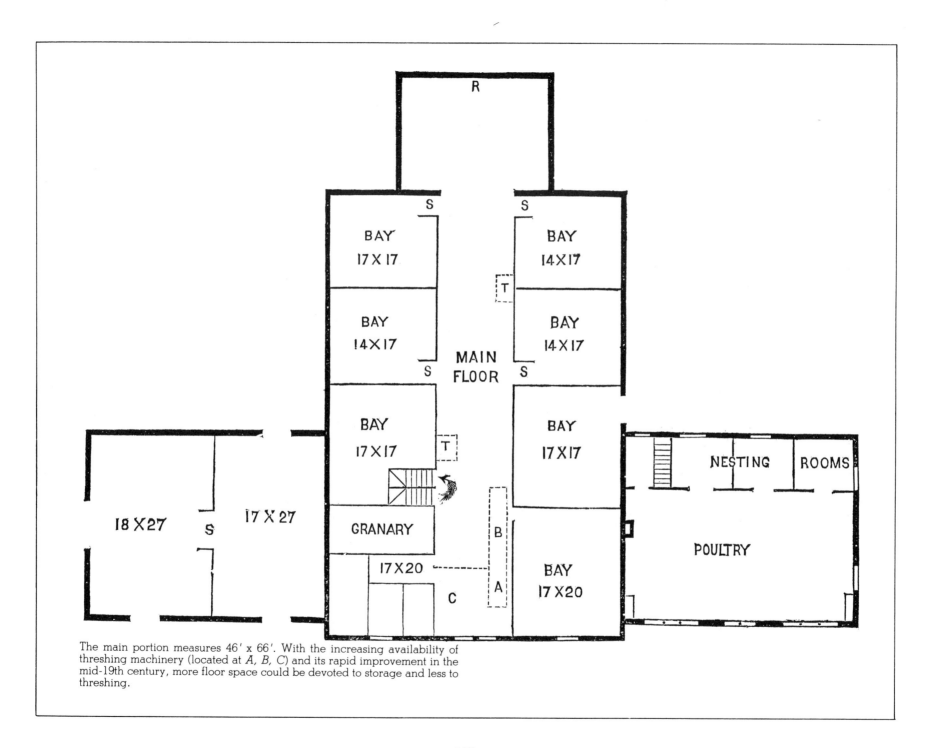

The main portion measures 46' x 66'. With the increasing availability of threshing machinery (located at *A*, *B*, *C*) and its rapid improvement in the mid-19th century, more floor space could be devoted to storage and less to threshing.

A Missouri barn, built in 1868, stands 86' square and contains 84 stalls for horses and cows. "It is regarded by well-informed people as the best barn in the State" for its architectural style and practical utility. Skylights between cupolas illuminate the mows.

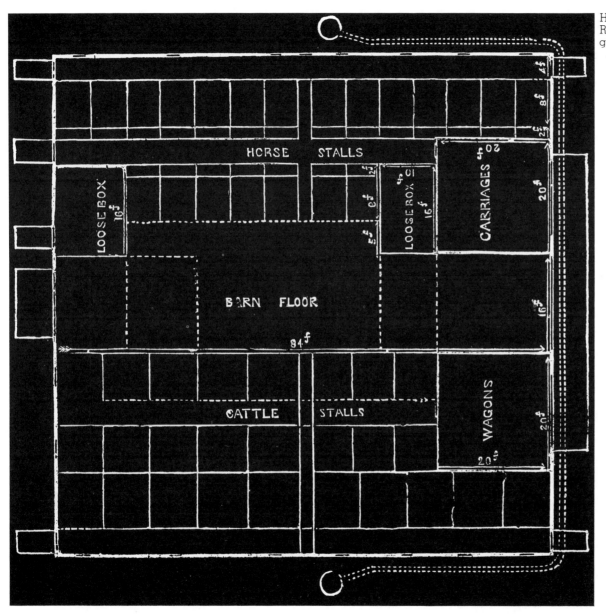

Hayforks travel along the interior dotted lines. Rainwater cisterns are connected by underground pipes along the outside walls.

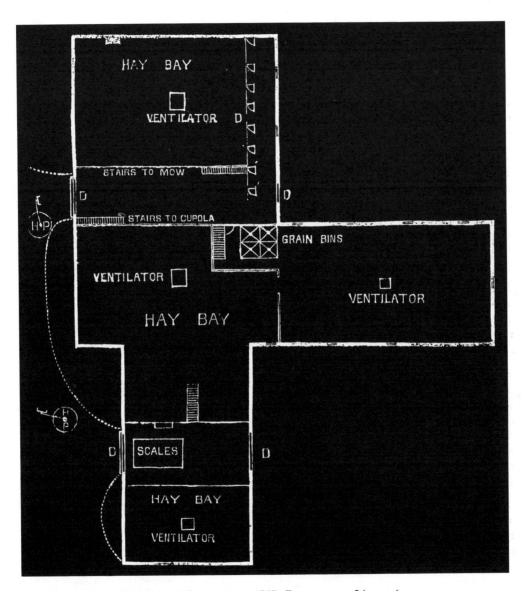

David Lyman barn, Middlefield, Connecticut, 1867. *Facing page:* A barn of this enormity—it covers one-fifth of an acre—can only have been built by a wealthy man; in this case, the builder was a Connecticut industrialist and farmer. *Above:* Plan of hay floor. Additional floor plans for this barn are on the following two pages.

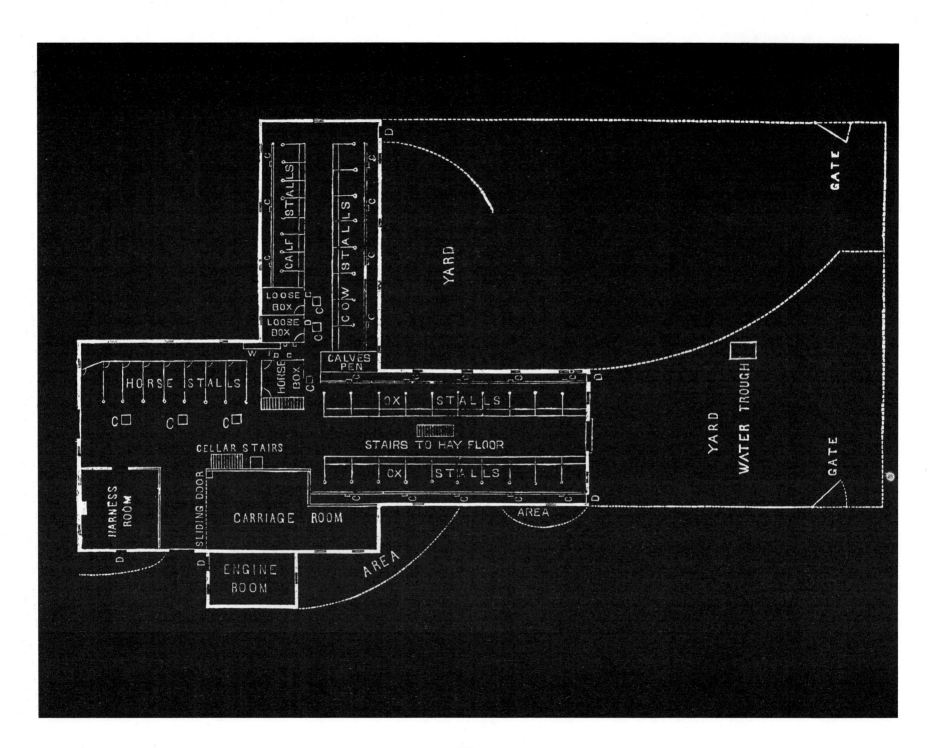

120

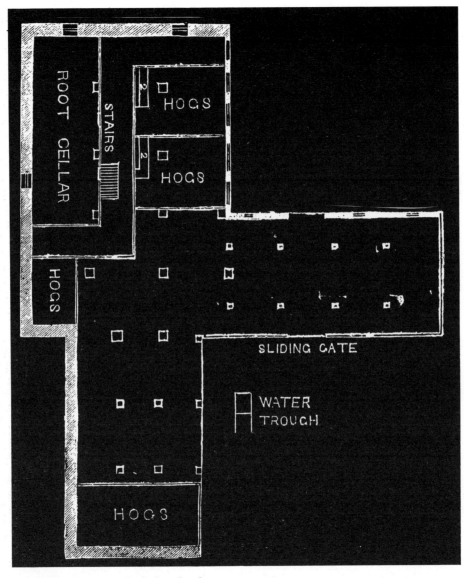

ROOT CELLAR

STAIRS

HOGS 2

HOGS 2

HOGS

HOGS

SLIDING GATE

WATER TROUGH

Facing page: Although the steam engine was not yet installed when this drawing was made, a fireproof room is appended to the main floor in anticipation of its purchase. *Above:* Cellar plan.

Stylistic similarities between domestic and farm architecture are clearly illustrated in the connected house, shed, and barn. Clipped gable, or "jerkinhead," roofs are slate-covered.

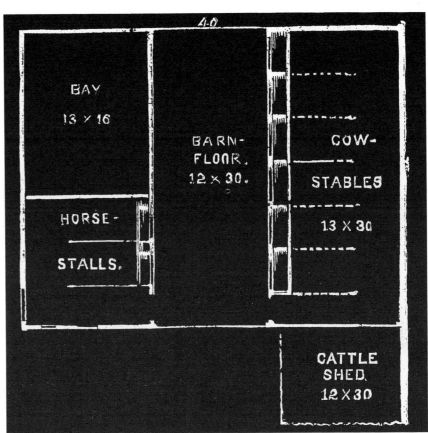

| BAY 13 × 16 | BARN-FLOOR. 12 × 30. | COW-STABLES 13 × 30 |
| HORSE-STALLS. | | CATTLE SHED. 12 × 30 |

Stables are located as far from the house as is possible. Hay is stored above the stable and in the bay on the ground floor.

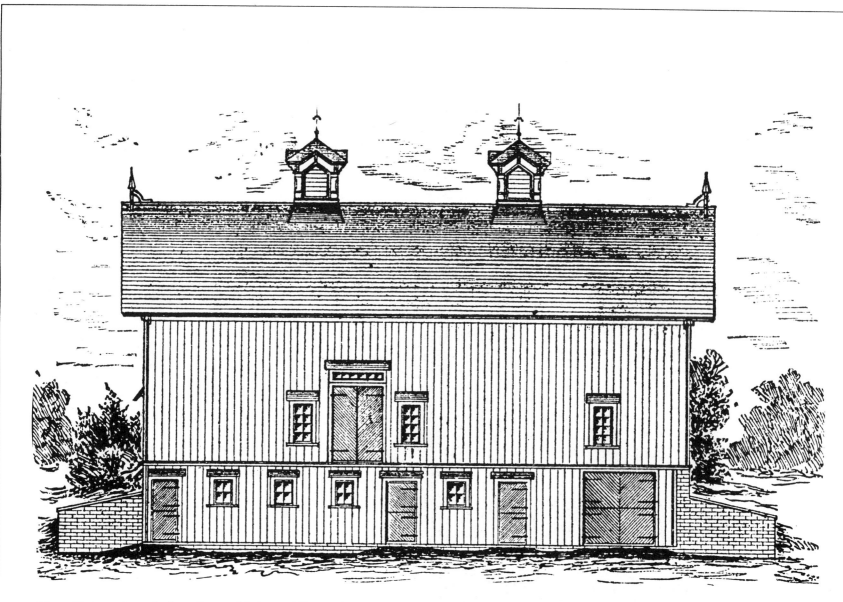

A large Victorian barn, Jackson County, Michigan, 1884, combines "convenience of arrangement with neatness of appearance," according to its designer, Edward Pratt.

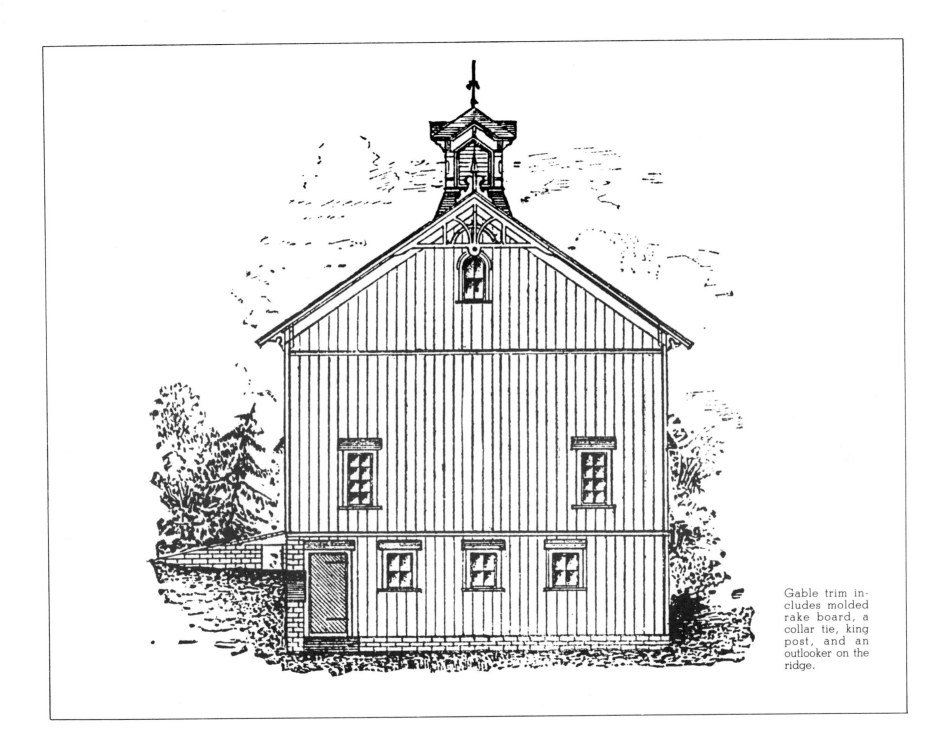

Gable trim includes molded rake board, a collar tie, king post, and an outlooker on the ridge.

125

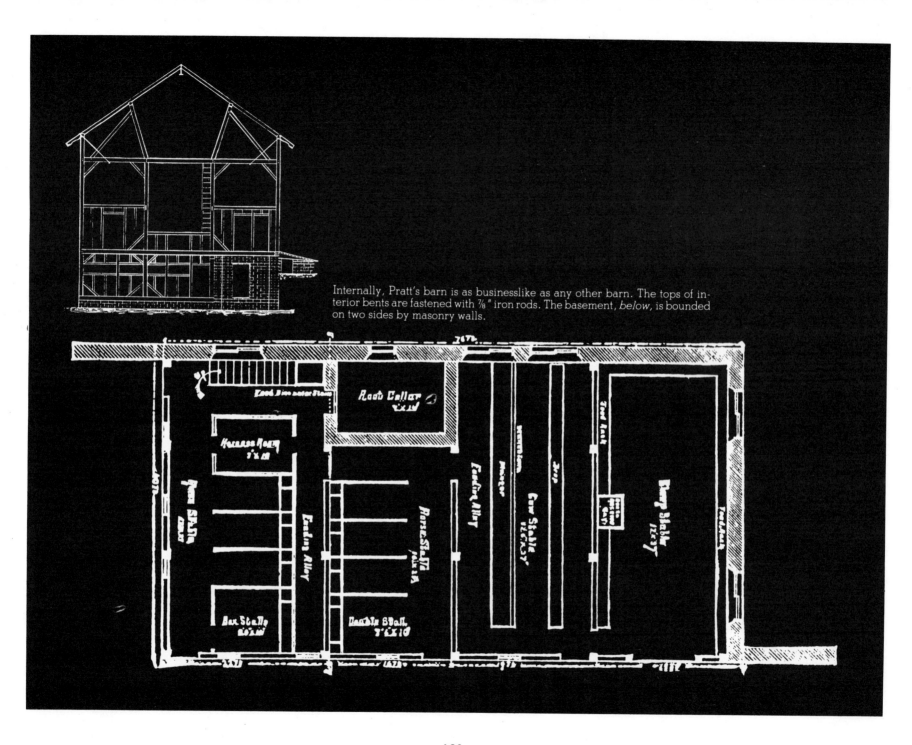

Internally, Pratt's barn is as businesslike as any other barn. The tops of interior bents are fastened with 7/8" iron rods. The basement, *below,* is bounded on two sides by masonry walls.

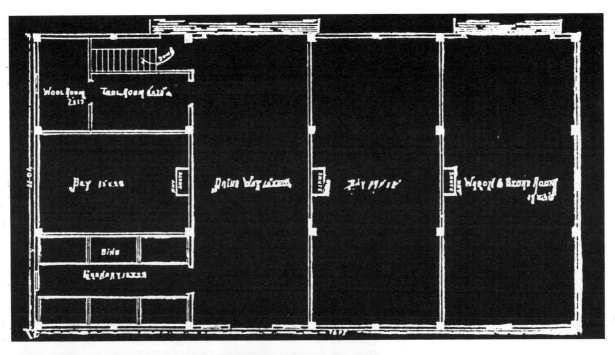

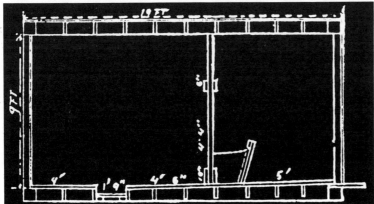

Above: Main floor plan of Pratt's barn. *Left:*
Section of cattle stalls.

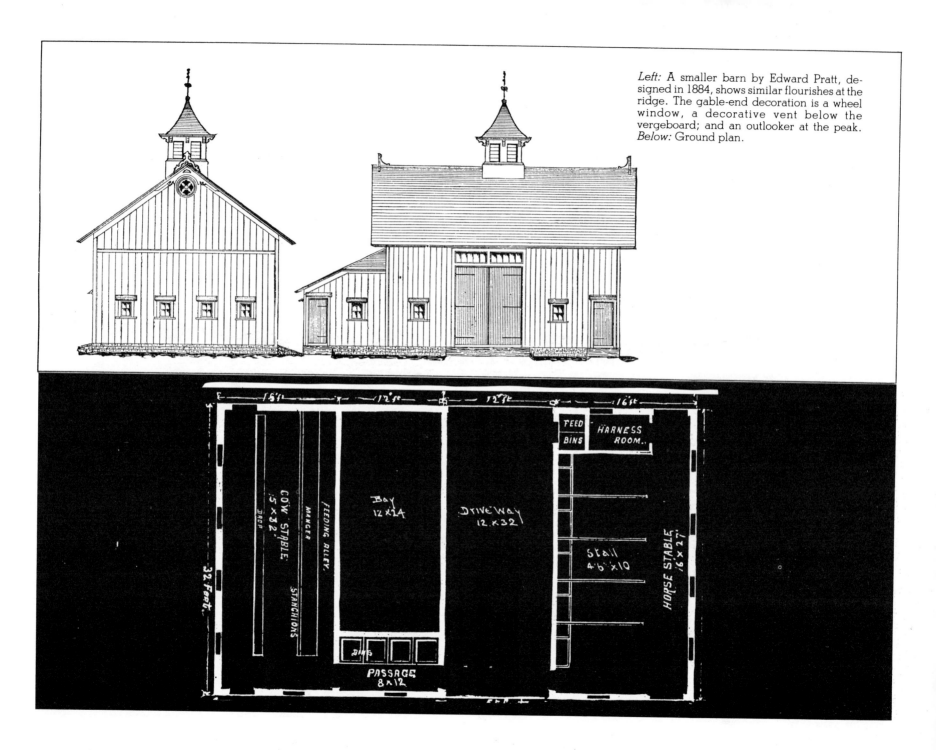

Left: A smaller barn by Edward Pratt, designed in 1884, shows similar flourishes at the ridge. The gable-end decoration is a wheel window, a decorative vent below the vergeboard; and an outlooker at the peak. *Below:* Ground plan.

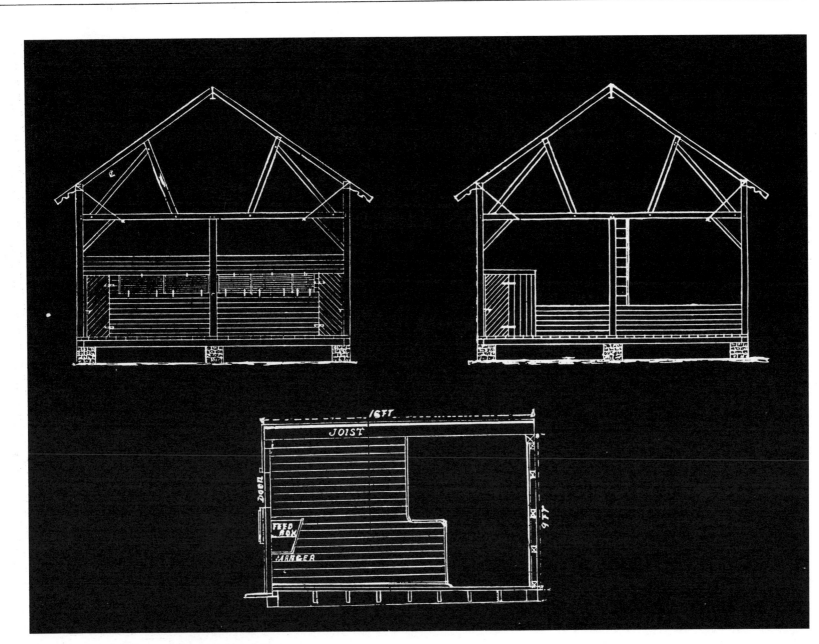

Clockwise from left: Section of horse stable; stable
side of driveway; bay side of driveway.

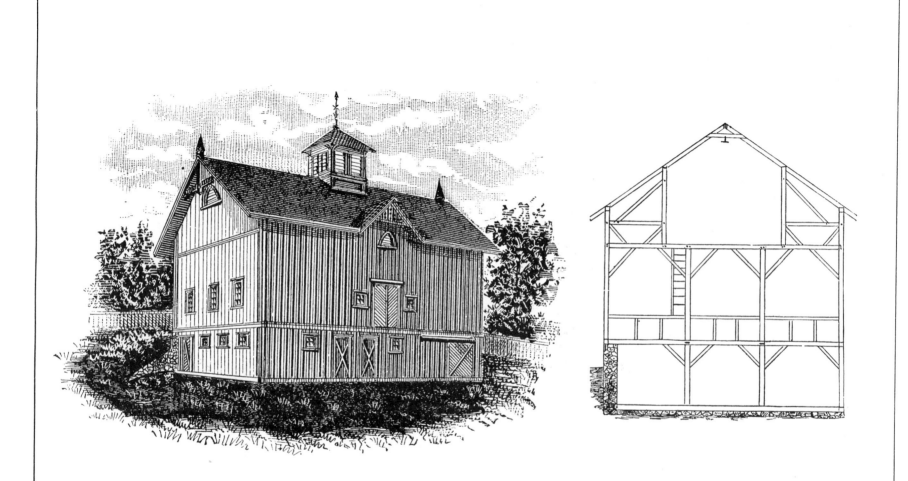

In an 1882 design by Edward Pratt, *above and facing page,* the roof line is broken with the addition of a gable over the doors on each side of the barn. *Above, right:* Absence of a tie beam allows for free movement of a hay fork.

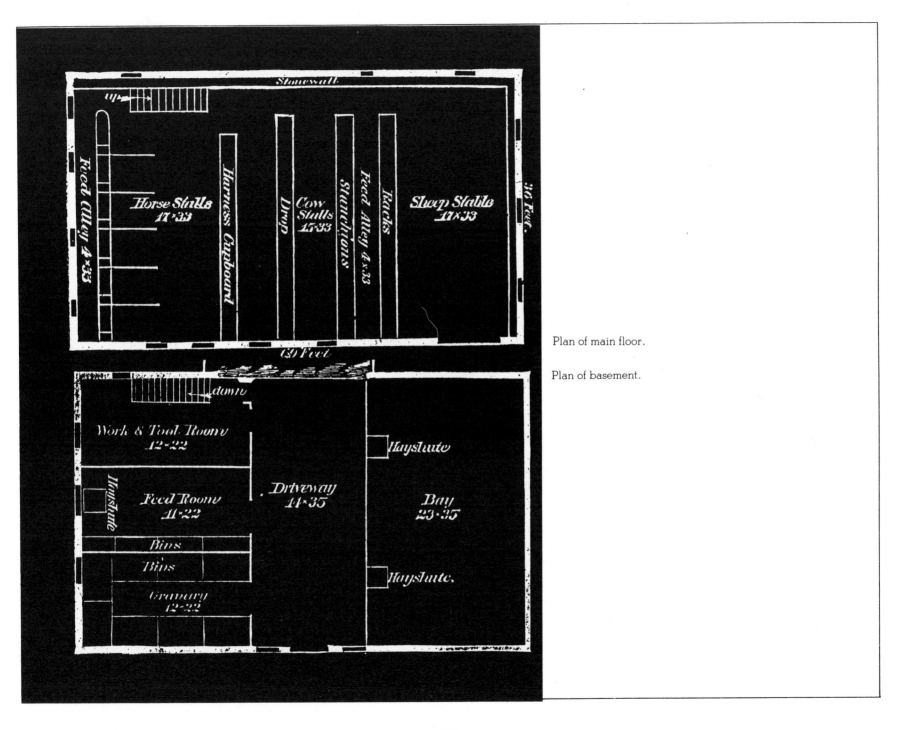

Plan of main floor.

Plan of basement.

In 1885, many barns continued to be built with pegged mortise and tenon joinery. Judge Benedict's barn, Richmond County, New York, *below and facing page,* claims the additional benefits of such modern conveniences as an 8-horsepower steam engine. It is located on the threshing floor and occupies less space than the 8 horses it theoretically replaces.

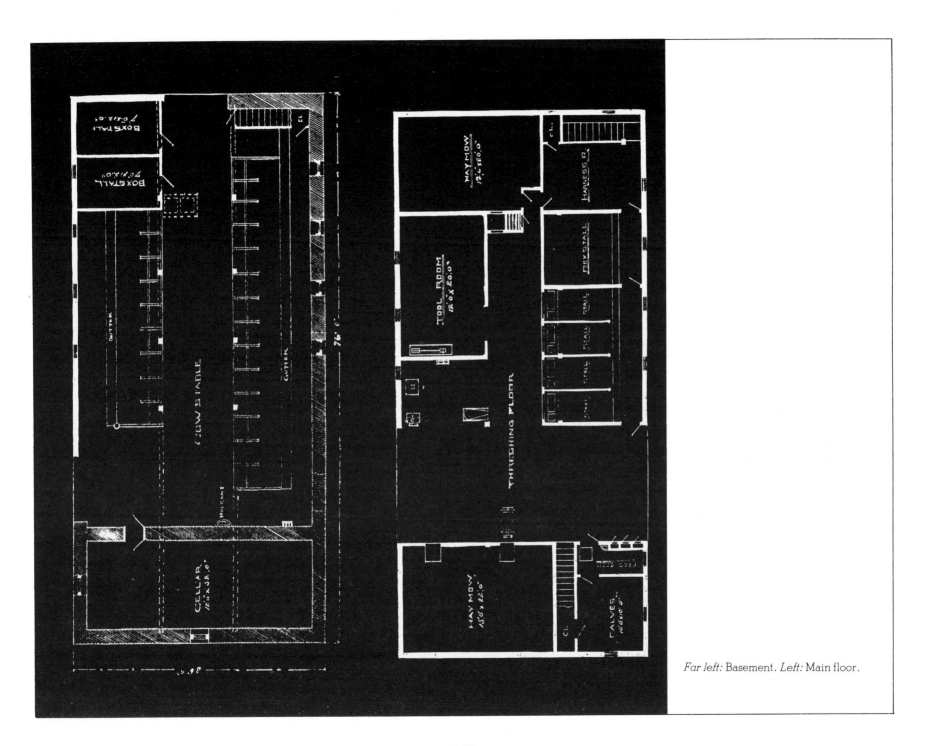

Far left: Basement. *Left:* Main floor.

The gable-end opening barn in Esdaile, Wisconsin, is fully mechanized by the standards of 1886. Horses still plowed the fields but did little work in the barn.

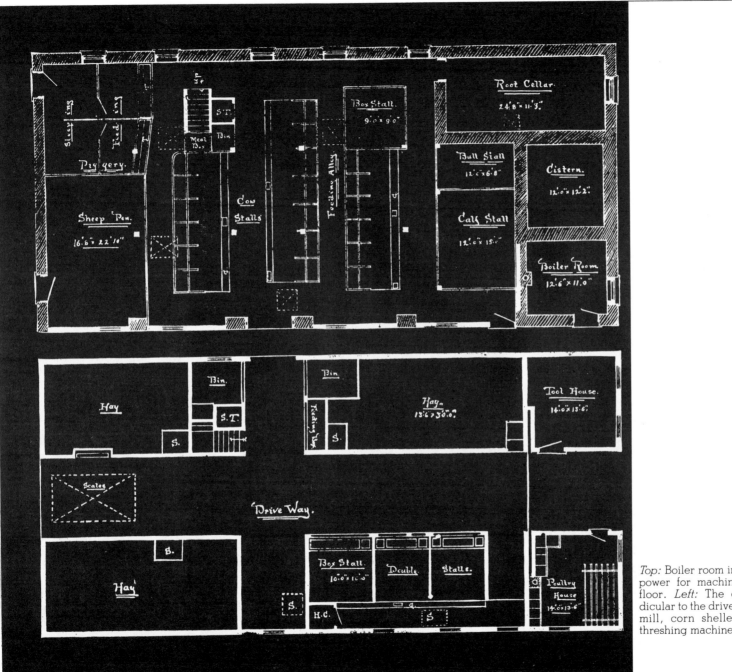

Top: Boiler room in the corner provides power for machines on the threshing floor. *Left:* The cross aisle, perpendicular to the driveway, contains a grist-mill, corn sheller, feed cutter, and threshing machine.

135

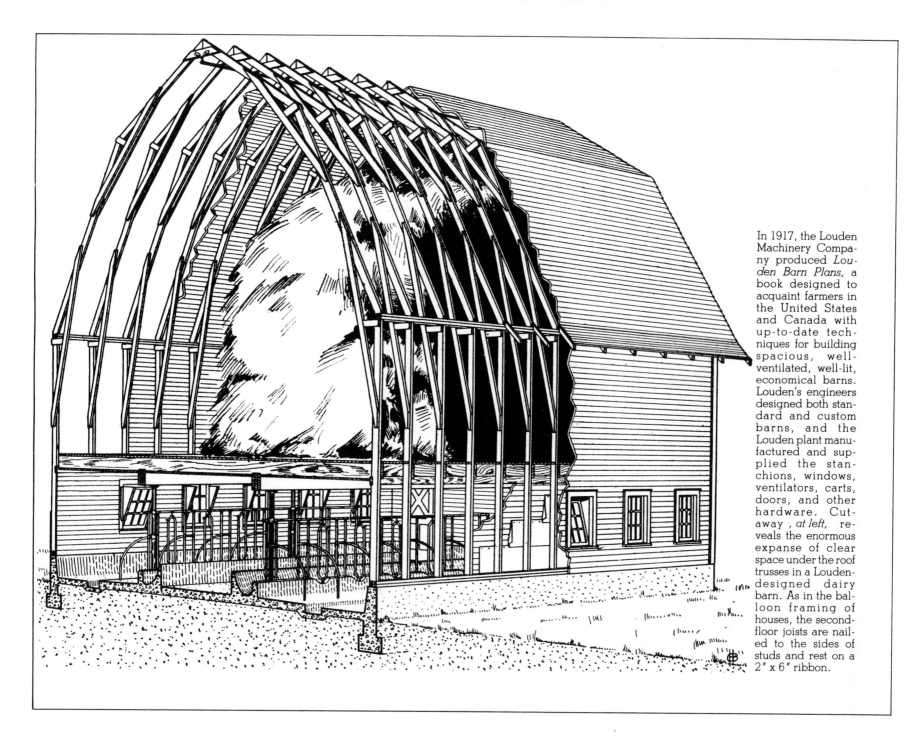

In 1917, the Louden Machinery Company produced *Louden Barn Plans,* a book designed to acquaint farmers in the United States and Canada with up-to-date techniques for building spacious, well-ventilated, well-lit, economical barns. Louden's engineers designed both standard and custom barns, and the Louden plant manufactured and supplied the stanchions, windows, ventilators, carts, doors, and other hardware. Cutaway , *at left,* reveals the enormous expanse of clear space under the roof trusses in a Louden-designed dairy barn. As in the balloon framing of houses, the second-floor joists are nailed to the sides of studs and rest on a 2" x 6" ribbon.

modern barns

The story of modern farm buildings is written in concrete and steel and brings to an end an era of great North American barn design. Because it was inexpensive, durable, and easy to clean, concrete was ideally suited for use in foundations and floors. Many states and provinces were soon to require it for sanitary reasons in dairies. Long in use in urban factories, galvanized iron rapidly gained in popularity as roofing and siding material for rural areas.

The evolution of scientific agriculture dovetailed nicely with advances in construction techniques and materials. Wood sill plates were bolted to the top of a foundation and light milled lumber (never thicker than 1 5/8 inches) was used to frame out the walls and roof. As the price of lumber increased, so did builders' knowledge of economy in the use of it. They knew that each piece should be only as large as was needed to safely withstand the strains to which it would be subject.

The continued concern with lighting and ventilation brought developments, too. Specially-designed windows took into account the need to minimize drafts and maximize the entry of sunlight. Elaborate venting systems were installed that operated on the principles of expansion and contraction of air at various temperatures. It would take another generation before the benefits of electricity reached most barns (or farmhouses, for that matter) in the United States and Canada.

Much of the great western plains territory of Canada at the turn of the century was only then being settled and put under cultivation. The Canadian Pacific Railroad encouraged settlement and commerce along its newly-extended transcontinental route by establishing loan programs and even providing detailed blueprints for building farmsteads in the West. Although simple in conception (as an inexpensive barn would have to be), these plans incorporated the same framing details seen in most modern barns in older rural areas.

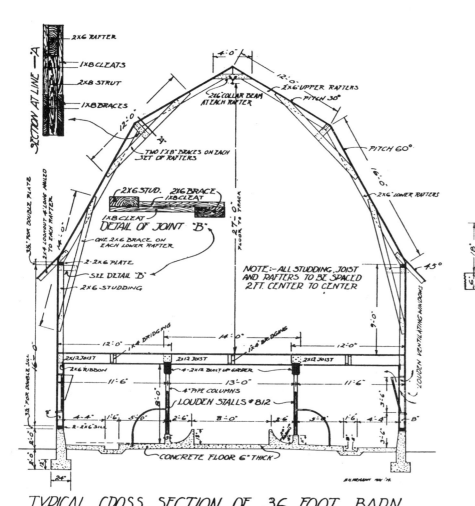

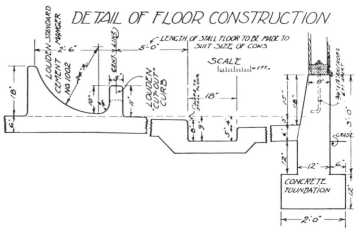

SECTION AT LINE "A"

2X6 RAFTER
1X8 CLEATS
2X8 STRUT
1X8 BRACES

2X6 COLLAR BEAM AT EACH RAFTER

4'-0"
12'-0"
2X6 UPPER RAFTERS
PITCH 30°

PITCH 60°

TWO 1"X8" BRACES ON EACH SET OF RAFTERS

2X6 STUD. 2X6 BRACE
1X8 CLEAT

1X8 CLEAT
DETAIL OF JOINT "B"

2X6 LOWER RAFTERS

ONE 2X6 BRACE ON EACH LOWER RAFTER

3½"OR DOUBLE PLATE

2X4 LOOKOUT 4' LONG NAILED TO EACH RAFTER

2-2X6 PLATE

SEE DETAIL "B"

2X6 STUDDING

NOTE:- ALL STUDDING, JOIST AND RAFTERS TO BE SPACED 2 FT. CENTER TO CENTER

45°

LOUDEN VENTILATING WINDOWS

12'-0"
1"X8 BRIDGING
14'-0"
1"X8 BRIDGING
12'-0"

2X12 JOIST
2X12 JOIST
2X12 JOIST

2X6 RIBBON
4-2X12 BUILT UP GIRDER
4" PIPE COLUMNS
LOUDEN STALLS #812

11'-6"
13'-0"
11'-6"

3½"OR DOUBLE SILL

2-2X6 SILL

4'-4" 1'-6" 5'-0" 2'-6" 8'-0" 2'-6" 5'-0" 1'-6" 4'-4" 8"

CONCRETE FLOOR 6" THICK

24"

TYPICAL CROSS SECTION OF 36 FOOT BARN

SCALE

ARCHITECTURAL DEPARTMENT
LOUDEN MACHINERY CO.
FAIRFIELD — IOWA.

DETAIL OF FLOOR CONSTRUCTION

LOUDEN STANDARD CEMENT MANGER NO. 1002

LENGTH OF STALL FLOOR TO BE MADE TO SUIT SIZE OF COWS

CENTER LINE

SCALE

LOUDEN CUT-OUT CURB

2'-6" 5'-0"

18"

CONCRETE FOUNDATION

GRADE

2'-0"

Compared with that of a traditional post and beam barn, the Louden dairy barn's braced rafter construction is easier to build and requires fewer hands to frame. The concrete floor is poured into forms of the desired shape.

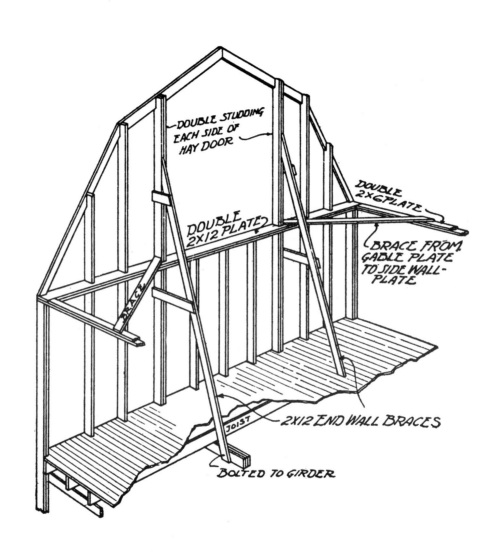

The gable ends require sufficient bracing against racking from wind and the pressure of the hay.

The plank truss method spaces the trusses farther apart (at intervals of 14' or 16') than the braced rafter type (intervals of 16" or 24"). The trusses support a purlin which in turn carries the individual rafters.

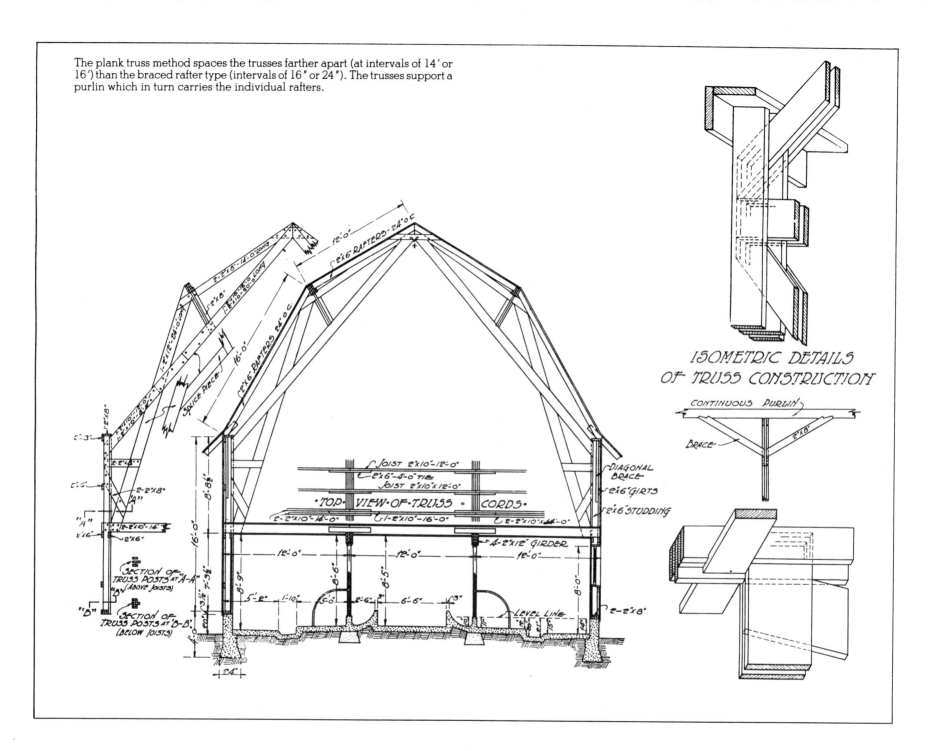

ISOMETRIC DETAILS
OF TRUSS CONSTRUCTION

CONTINUOUS PURLIN

BRACE

TOP·VIEW·OF·TRUSS·CORDS·

SECTION OF
TRUSS POSTS AT "A-A"
(ABOVE JOISTS)

SECTION OF
TRUSS POSTS AT "B-B"
(BELOW JOISTS)

140

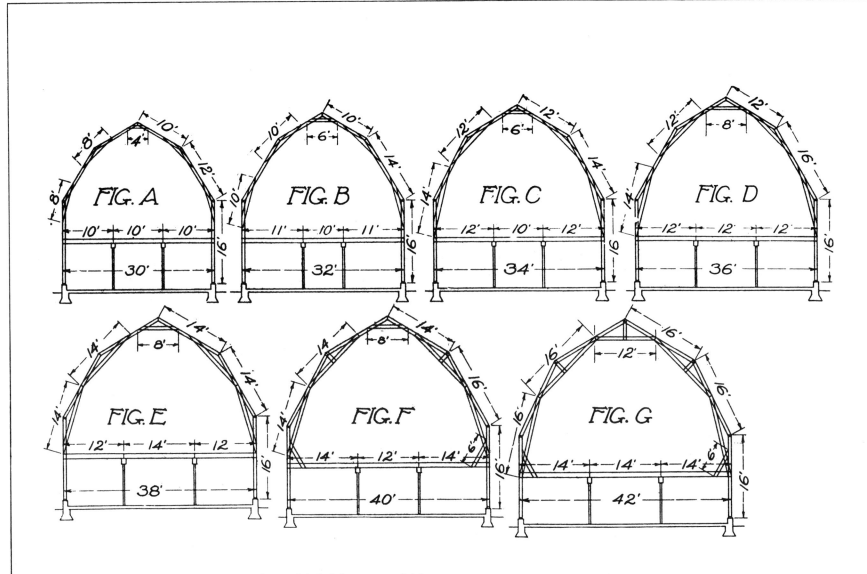

FIG. A   FIG. B   FIG. C   FIG. D

FIG. E   FIG. F   FIG. G

Milled lumber, even in 1917, was smaller than its labeled dimensions. A 2" x 6" beam, for example, actually measured 1 5/8" x 5 5/8". In order to get the most strength out of the least number of board feet, Louden's engineers devised a series of recommended dimensions for framing out roof spans of various typical widths.

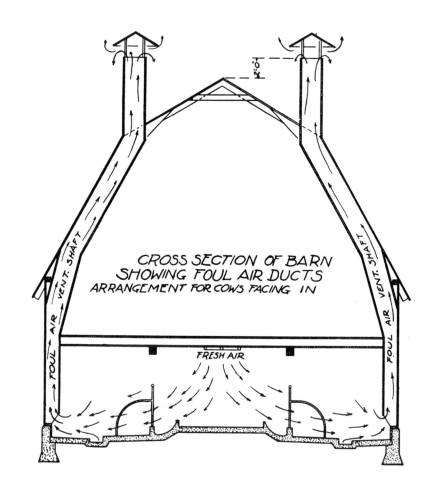

CROSS SECTION OF BARN
SHOWING FOUL AIR DUCTS
ARRANGEMENT FOR COWS FACING IN

FOUL AIR VENT. SHAFT

FOUL AIR VENT. SHAFT

FRESH AIR

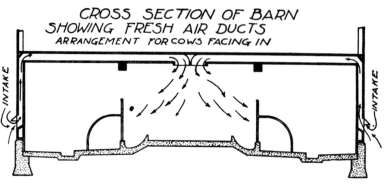

CROSS SECTION OF BARN
SHOWING FRESH AIR DUCTS
ARRANGEMENT FOR COWS FACING IN

INTAKE

INTAKE

A healthy dairy requires fresh air. In one Louden design, colder outside air enters intakes and forces warmer air out vents in the room. Ventilation does not work when the outside and inside temperatures are equal.

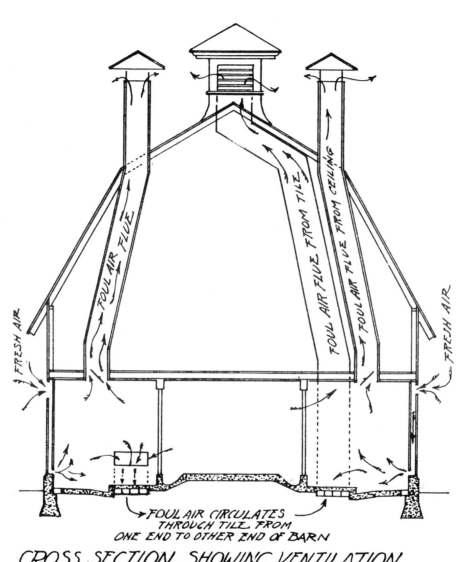

FRESH AIR

FOUL AIR FLUE

FOUL AIR FLUE FROM TILE

FOUL AIR FLUE FROM CEILING

FRESH AIR

FOUL AIR CIRCULATES
THROUGH TILE FROM
ONE END TO OTHER END OF BARN

CROSS SECTION SHOWING VENTILATION

Another system of ventilation, *left,* and in detail, *overleaf.*

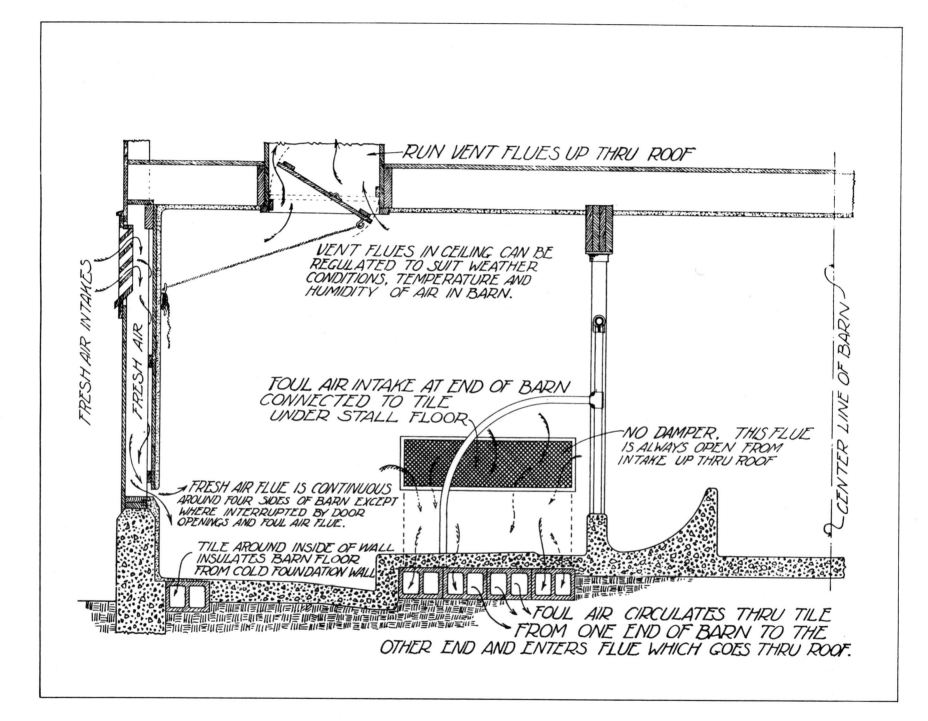

RUN VENT FLUES UP THRU ROOF

FRESH AIR INTAKES

FRESH AIR

VENT FLUES IN CEILING CAN BE REGULATED TO SUIT WEATHER CONDITIONS, TEMPERATURE AND HUMIDITY OF AIR IN BARN.

FOUL AIR INTAKE AT END OF BARN CONNECTED TO TILE UNDER STALL FLOOR.

NO DAMPER. THIS FLUE IS ALWAYS OPEN FROM INTAKE UP THRU ROOF

CENTER LINE OF BARN

FRESH AIR FLUE IS CONTINUOUS AROUND FOUR SIDES OF BARN EXCEPT WHERE INTERRUPTED BY DOOR OPENINGS AND FOUL AIR FLUE.

TILE AROUND INSIDE OF WALL INSULATES BARN FLOOR FROM COLD FOUNDATION WALL

FOUL AIR CIRCULATES THRU TILE FROM ONE END OF BARN TO THE OTHER END AND ENTERS FLUE WHICH GOES THRU ROOF.

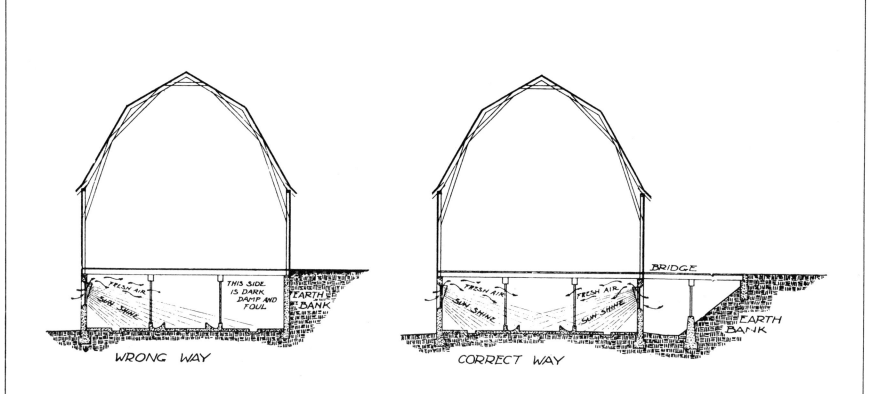

WRONG WAY

THIS SIDE IS DARK DAMP AND FOUL

FRESH AIR

SUN SHINE

EARTH BANK

CORRECT WAY

BRIDGE

FRESH AIR

SUN SHINE

FRESH AIR

SUN SHINE

EARTH BANK

Adequate light is as important as fresh air, especially in the basement of a bank barn. Farmers of the late 19th and early 20th centuries were encouraged to grade a passageway along the uphill side and build a bridge to the main floor and to install windows in the exposed wall.

145

Below and *opposite page:* Long, low dairy barn and separate hay shed reflect the early 20th-century trend toward specialized farm buildings. The dairy is compact, well-insulated, and cheaper than if it had a second floor. The hay shed is much simpler to put up than a two-story dairy barn and it holds more hay.

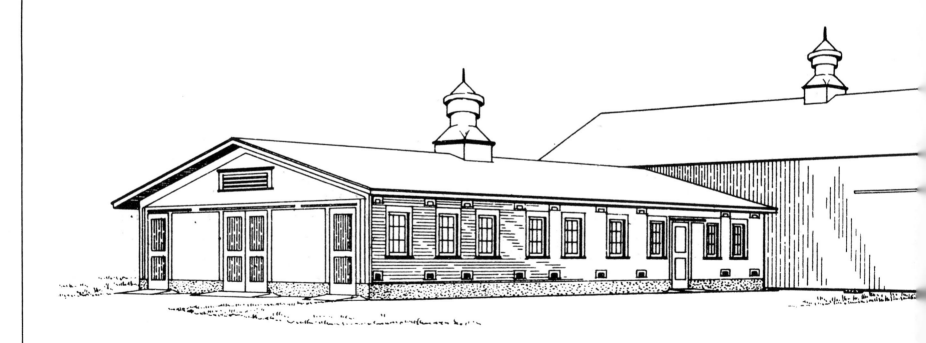

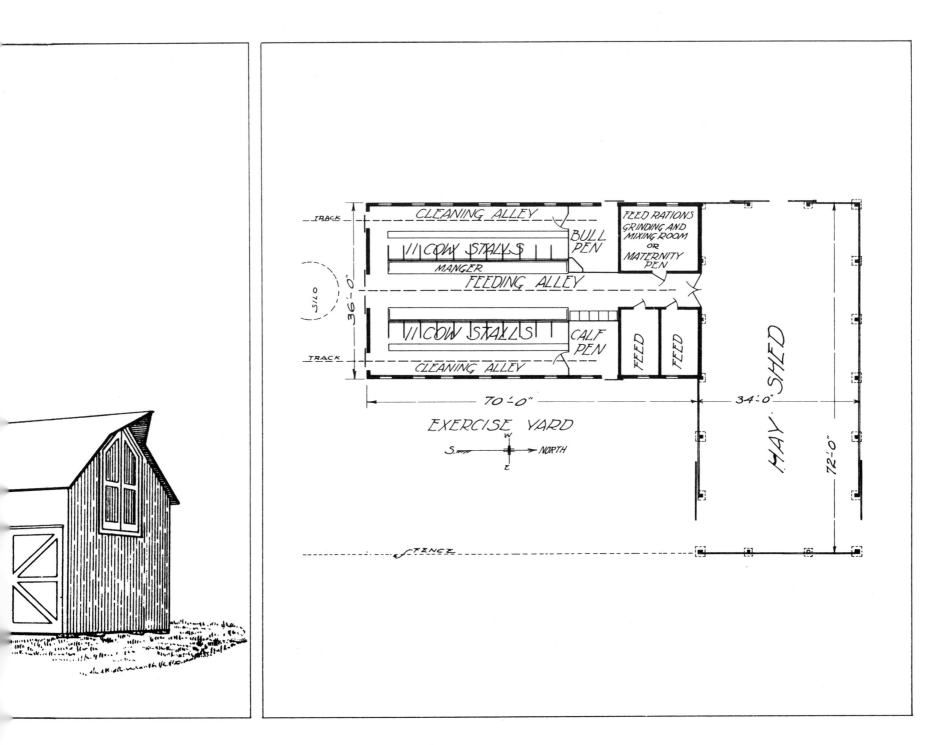

TRACK

CLEANING ALLEY

BULL
PEN

FEED RATIONS
GRINDING AND
MIXING ROOM
OR
MATERNITY
PEN

VII COW STALLS

MANGER

FEEDING ALLEY

36'-0"

SILO

VII COW STALLS

CALF
PEN

FEED

FEED

TRACK

CLEANING ALLEY

70'-0"

34'-0"

HAY SHED

72'-0"

EXERCISE YARD

S. ← W → NORTH
E

FENCE

147

In this Louden barn, the foundation wall and the entire floor are built of concrete while the barn itself is of plank-frame construction. An overhead track runs from the silo along the feed alleys. Manure is easily collected by running a cart along the passageway between stalls.

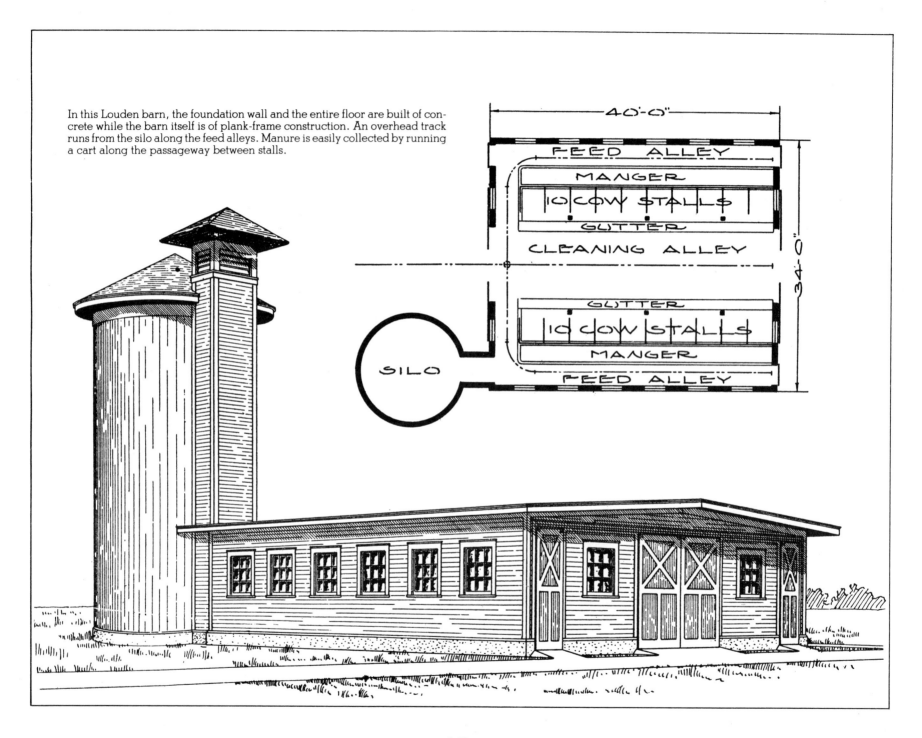

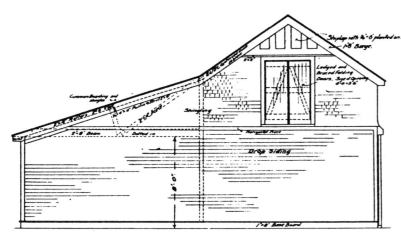

- END ELEVATION -

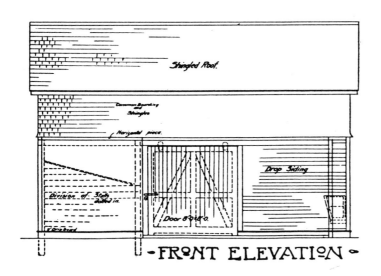

- FRONT ELEVATION -

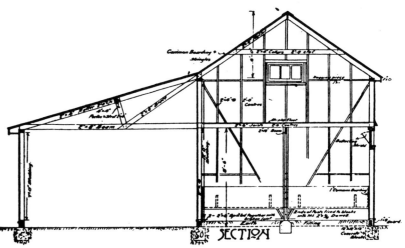

SECTION

Common milled lumber could be used to build the following modest barns (pages 149-154) designed by the Canadian Pacific Railroad's Department of Natural Resources in 1913. Posts sunk into concrete blocks eliminated the need to dig footings or to build a foundation. Framing required 2″ x 4″ studs and rafters (24″ on center), 2″ x 6″ sills and joists, and 2″ x 8″ and 2″ x 10″ beams.

A farmer could accommodate a small herd of cattle in the stable and in the covered, but open shed, and an adequate supply of hay and bedding in the loft.

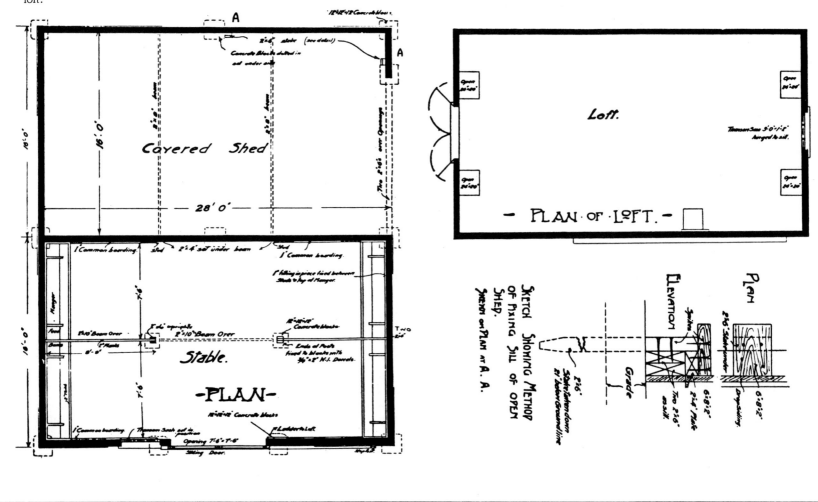

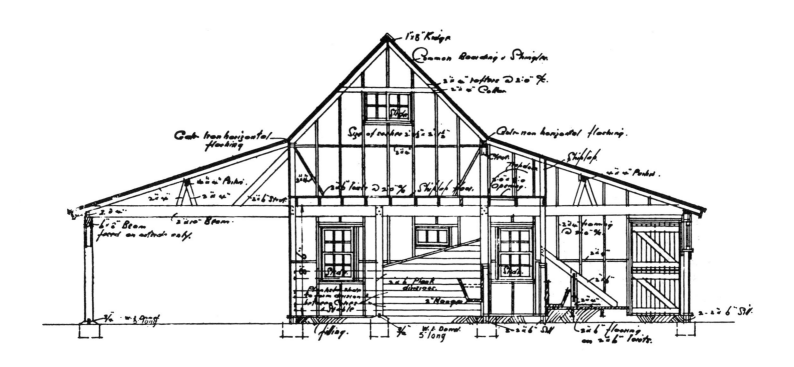

The barn illustrated on pages 151-154 provides considerably more space for stock and storage than did the previous plan. Like the first, on pages 149-150, it is anchored in blocks and is designed to conserve both material and labor. Its roof, although wide and not steep, could withstand wind and snow loading.

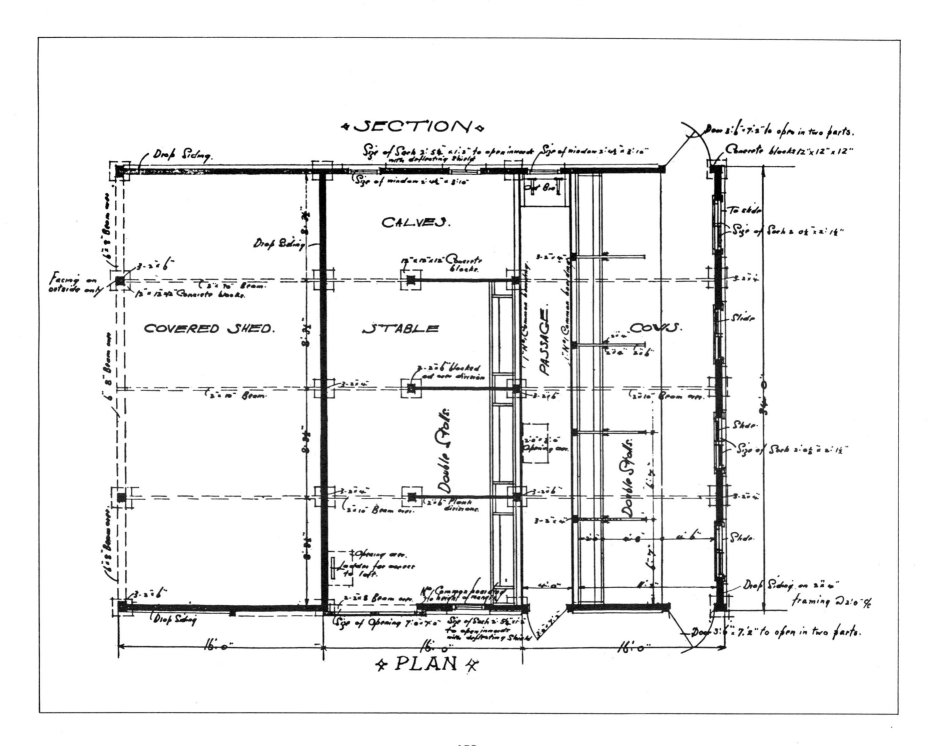

‹ SECTION ›

✦ PLAN ✦

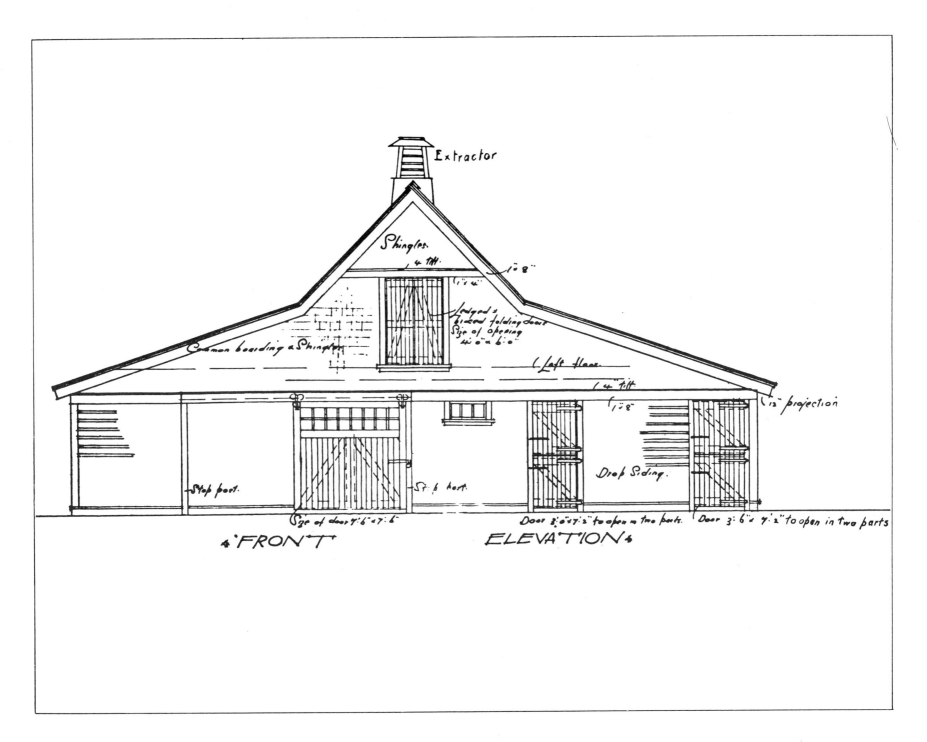

Extractor

Shingles.
4 7H.

1'-8"

Common boarding & Shingles.

Hinged & slide folding doors
Size of opening
4'-6" x 6'-0"

Loft floor.

4 7H.

1'-8"

12" projection

Stop post.

Drop Siding.

Size of door 7'-6" x 7'-6"

St. 6 post.

Door 3'-0" x 7'-2" to open in two parts.

Door 3'-6" x 7'-2" to open in two parts

4 FRONT   ELEVATION 4

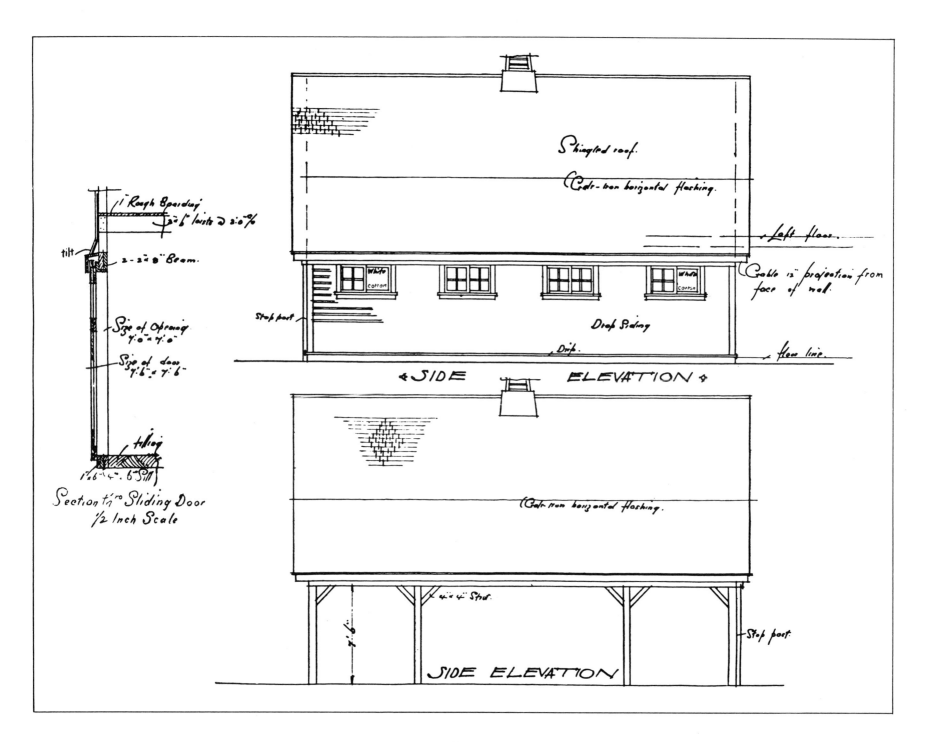

Section thro Sliding Door
½ Inch Scale

SIDE ELEVATION

SIDE ELEVATION

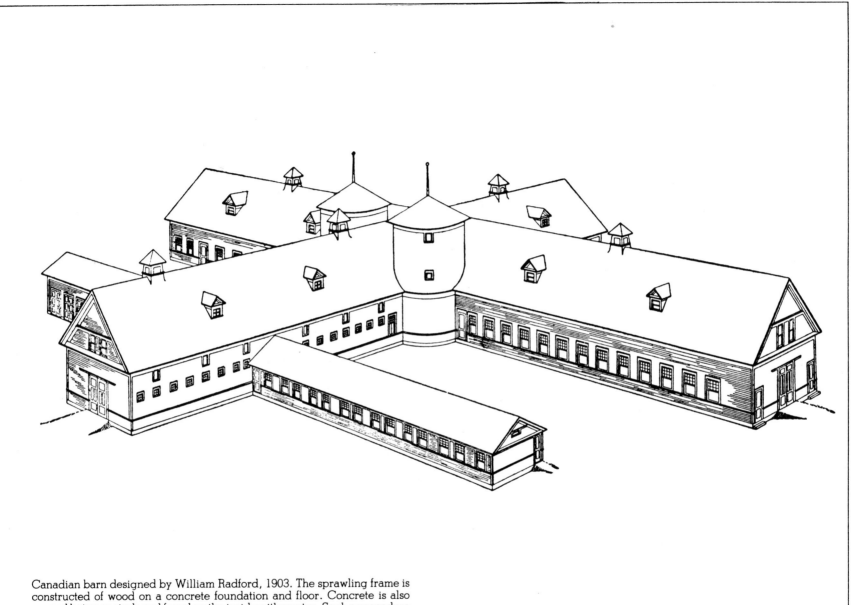

Canadian barn designed by William Radford, 1903. The sprawling frame is constructed of wood on a concrete foundation and floor. Concrete is also poured between studs and faced on the inside with mortar. Such a procedure makes for easy cleaning but poor insulation as concrete is a good conductor of cold. An additional elevation and a floor plan are shown on pages 156-157.

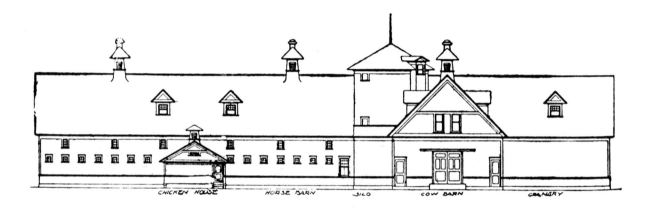

CHICKEN HOUSE   HORSE BARN   SILO   COW BARN   GRANARY

A granary is situated at the center of the north side and contains large bins connected to a mechanical conveyor. The main driveway passes through the granary. In the cow barn, 57 head are fed from a suspended trolley system. Manure gutters drain all liquids into a single, central catch basin. Air ducts connect to ventilators on the roof. A separate wing contains stalls for young stock and an isolated hospital area. A chicken house, horse barn, and equipment shed are also under the same connected roof. The centrally-located silos are studded and sheathed on the inside and lined with vitreous paving brick. The capacity of each silo is about 50 tons.

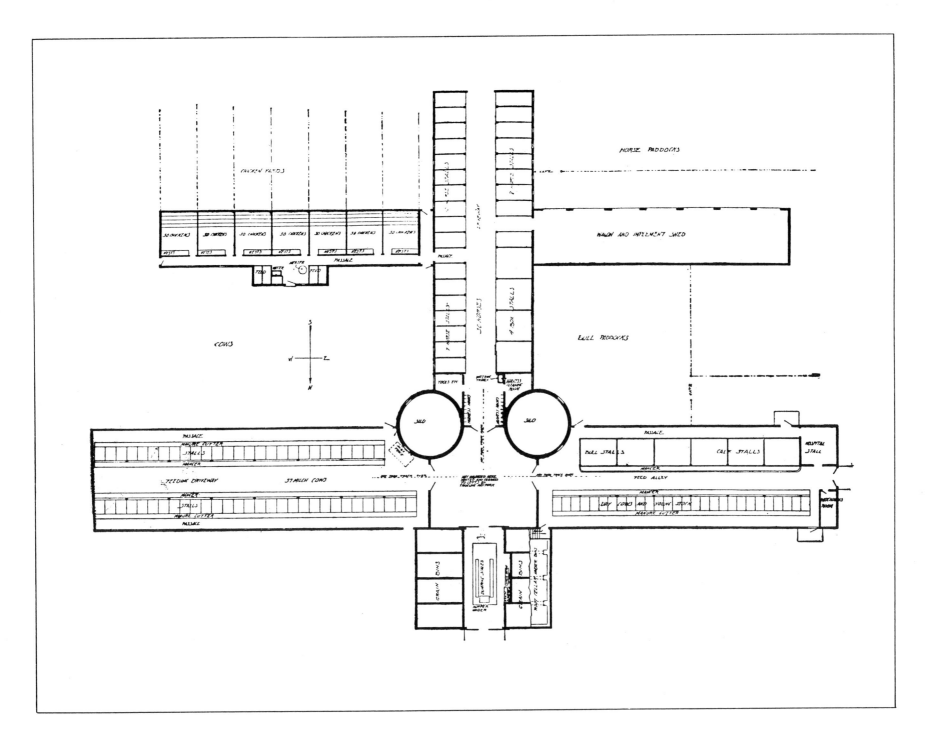

glossary

ADZE. An axe-like tool with an arched blade used to dress timbers.

ANCHORBEAM. The massive crossbeam found in Dutch barns.

BALLOON FRAMING. A type of framing in which the studs continue from the sill to the base of the rafters without a beam, or girt, for second-floor joists.

BANK BARN. A barn built into a hillside or against a rise of earth allowing ground-level entry to the upper floor from the uphill side and to the basement from the downhill side. Also known as a sidehill or basement barn.

BAY. A compartment in a barn usually divided from another bay by vertical posts and a horizontal beam.

BENT. A cross-sectional unit of framing consisting of at least two columns connected by a beam, usually assembled on the ground and then swung up into place where it is joined to longitudinal members.

BROADAXE. An axe with a wide, flat head and a short handle used in squaring logs.

CHINKING. Material used to fill cracks between logs; See also WATTLE AND DAUB.

CLAPBOARD. Overlapped, horizontal exterior siding.

COLLAR BEAM. A horizontal member tying together two common rafters.

FROE. A hand tool for splitting shingles; also spelled *frow.*

GABLE. The wall at the end of a ridged roof.

GAMBREL. A ridged roof having two slopes on each side, the lower slope having the steeper pitch.

GIRT. A horizontal member between columns or bents. It acts as a stiffener and is used to support rafter ends.

GOTHIC ROOF. A pointed-arch roof.

HORSE POWER.   A wheel or combination of gears to which a horse is harnessed, the turning of which operates various farm machines.

MORTISE.   A cutout in a log or beam made to receive the tenon of another to which it is to be joined.

MOW.   A place for storing hay or grain.

PENT ROOF.   A roof in one sloping plane.

POST AND BEAM FRAMING.   A type of framing in which horizontal members rest on a post.

PURLIN.   A horizontal member usually supporting rafters at mid-span.

RIGID ARCH CONSTRUCTION.   A modern building type providing unobstructed space from floor to ceiling.

TENON.   A projection in a log or beam made to fit into the mortise of another to form a joint.

THRESHING FLOOR.   A wide, flat area in a barn used as the site for separating grain from straw. It became obsolete after the introduction of the combine, which in one operation cut and threshed crops in the field.

TRUSS.   An arrangement of straight timbers designed to distribute stresses across a span.

WATTLE AND DAUB.   Interwoven twigs packed with mud or plaster and pressed into cracks between logs to improve sheltering.